Foundations of Professional Psychology

Melchert's book is a valuable resource for graduate students and their faculty to understand how professional psychology is transforming into a health profession, while maintaining its unique psychological identity. I highly recommend it.

Dr. James Bray
Past President, American Psychological Association
Editor, *Primary Care Psychology* and *Handbook of Family Psychology*

Foundations of Professional Psychology
The End of Theoretical Orientations and the Emergence of the Biopsychosocial Approach

Timothy P. Melchert
Marquette University
Milwaukee, WI, USA

AMSTERDAM • BOSTON • HEIDELBERG • LONDON • NEW YORK • OXFORD
PARIS • SAN DIEGO • SAN FRANCISCO • SINGAPORE • SYDNEY • TOKYO

Elsevier
32 Jamestown Road London NW1 7BY
225 Wyman Street, Waltham, MA 02451, USA

First edition 2011

Notices
Knowledge and best practice in this field are constantly changing. As new research and
experience broaden our understanding, changes in research methods, professional practices,
or medical treatment may become necessary.

Practitioners and researchers must always rely on their own experience and knowledge in
evaluating and using any information, methods, compounds, or experiments described
herein. In using such information or methods they should be mindful of their own safety and
the safety of others, including parties for whom they have a professional responsibility.

To the fullest extent of the law, neither the Publisher nor the authors, contributors, or editors,
assume any liability for any injury and/or damage to persons or property as a matter of
products liability, negligence or otherwise, or from any use or operation of any methods,
products, instructions, or ideas contained in the material herein.

British Library Cataloguing-in-Publication Data
A catalogue record for this book is available from the British Library

Library of Congress Cataloging-in-Publication Data
A catalog record for this book is available from the Library of Congress

ISBN: 978-0-12-385079-9

For information on all Elsevier publications
visit our website at elsevierdirect.com

This book has been manufactured using Print On Demand technology. Each copy is
produced to order and is limited to black ink. The online version of this book will show
color figures where appropriate.

Contents

Preface

Psychology has been remarkably successful both as a scholarly discipline and a field of clinical practice. It has had a transformative effect on important areas in health care, the social sciences, education, public policy, business, organizations, and even culture in general. As a result, society's understanding of mental health and psychological functioning has evolved dramatically and mental health treatment has provided relief to countless individuals.

Despite all of its remarkable successes across the research and practice domains, the growth and development of the field have also been quite complicated and contentious. Throughout its history, there has been major competition and conflict between theoretical camps and schools of thought. Though the conflict has subsided in recent years, students, faculty, and clinicians are certainly all familiar with the deep divides that have existed, and continue to exist, between adherents of different theoretical orientations, researchers and practitioners, qualitatively and quantitatively oriented researchers, and between adherents of the different approaches to education and practice.

This book addresses why the field developed in this complicated manner and whether it is now possible to resolve and move beyond the conflicts and competition of the past by adopting a fundamentally different approach to conceptualizing education and practice in the profession. The book examines the basic theoretical and conceptual foundations underlying professional psychology and evaluates the adequacy of those foundational frameworks in light of current scientific evidence and health care practices. It concludes that several important educational and clinical practices are outdated and that the basic conceptual foundations of the field need to be revised in order to be consistent with current scientific knowledge regarding human development, functioning, and behavior change. Updating these conceptual frameworks will help to resolve many long-standing conflicts as well as provide a coherent, unified perspective for moving the field forward.

The basic motivation for writing this book rose out of frustrations I have had ever since entering graduate school. During my coursework and when I applied for practicum placements, internship, and my first faculty position, I was routinely asked about my theoretical orientation to practice in the field. Though I could see value in being asked to explain my personal views on the nature of personality, psychopathology, and psychotherapy, it seemed obvious that these were impossibly difficult questions to answer. Far more intelligent and learned people than I have attempted to describe the nature of human psychology, and clearly their answers were controversial and were often viewed as seriously flawed. The various theories developed by Freud, Watson, Rogers, Ellis, more recent theorists, or biologically oriented psychiatrists, all appeared to be limited and were often harshly criticized or even ridiculed by those who endorsed competing orientations. In light of all this criticism, it seemed unlikely that *any* of the theoretical orientations was scientifically valid. Of course, professional psychology was supposed to be based on science, and yet one had to assume that science would not support multiple competing theories for understanding psychological phenomena. Nonetheless, I was expected to pick one of these as my personal orientation. Some professors suggested that if I found the existing orientations lacking, I could modify or combine them, or even come up with a new one of my own. In addition, new theoretical orientations were being developed and published regularly. This obviously was not an easy situation for a graduate student to navigate. There has been progress since that time in resolving aspects of this problem, but students today are still being asked these same types of questions. For instance, the uniform *Application for Psychology Internships* that is used by almost all of the psychology internship programs listed by the Association of Psychology Postdoctoral and Internship Centers (APPIC) requires applicants to complete the following essay item: "Please describe your theoretical orientation and how this influences your approach to case conceptualization and intervention" (APPIC, 2009, p. 22).

There is, however, a very solid science-based solution to this problem. This book examines problems like the ones described above in light of current scientific knowledge and health care practice, and concludes that several of the theoretical underpinnings of the field need to be updated. The book then presents a comprehensive solution to these problems. It does not propose another theoretical orientation, at least not in the traditional sense that that term has been used in the field. Instead, it argues for a comprehensive metatheoretical framework that is firmly grounded in the scientific understanding of human development, functioning, and behavior change. Though many aspects of psychology are still far from being understood in detail, the scientific understanding of psychology has advanced to the point where it can now support a unified biopsychosocial approach. This approach has already begun replacing traditional practices in a variety of areas within professional psychology, but there has not yet been a comprehensive description of how this approach applies for the field as a whole. After discussing the nature of a unified biopsychosocial approach to professional psychology, the book goes on to discuss the main implications of this approach for education, practice, and research.

This book addresses the field of professional psychology as a whole. Despite the different sets of knowledge and skills needed by psychologists working in the various general and specialized areas of practice in the field, there is a critical need for a unified science-based foundational framework that applies to all populations, all disorders, all behavior, and the whole field of professional psychology.

The book also takes a careful, systematic approach to analyzing the conceptual foundations of the field. It begins by examining the lack of clarity in the basic definitions used to identify the nature and purpose of professional psychology. It then goes on to identify the main reasons for the confusing theoretical landscape within the field and the development of the many competing theoretical orientations for understanding personality, psychopathology, and psychotherapy. After evaluating the requirements of a satisfactory solution to these problems, the book then goes on to show how a unified, biopsychosocial approach meets those requirements and applies to all human behavior and across all the general and specialized areas of professional practice. The book also shows how the biopsychosocial approach can be applied across the whole behavioral health care treatment process from intake assessment through the evaluation of the outcomes of treatment.

A primary audience for this book includes professional psychology graduate students and their faculty. The book advocates for a major updating of the conceptual frameworks that are used for structuring and organizing education, practice, and research in the field, issues that are centrally important in graduate education. The book can also be used in master's programs and with advanced undergraduate psychology majors to address the same questions. The book is also intended for established practitioners who are very experienced and familiar with the confusing conceptual landscape of the profession.

I am grateful to the many individuals who assisted me with the development of the ideas in this book and in preparing the text. Wrestling with the problems addressed in this book started many years ago in my own graduate education, most notably with my first doctoral advisor, Kent Burnett. It continued with my faculty colleagues over the years, and especially Todd Campbell, who shared with me the responsibility for directing all of our master's and doctoral training programs for many years. Lari Meyer completed her dissertation research under my supervision and her research figures prominently in Chapter 8. My wife, Susan Schroeder, is a highly skilled therapist who provides an excellent model for practicing psychotherapy in an integrated biopsychosocial manner. I also thank Lee Hildebrand, Augustine Kalemeera, Rebecca Mayor, Robert Nohr, and Lucia Stubbs, who shared their critiques of the book. I am indebted to all of these very fine individuals.

Part I

Introduction

This part of the book is comprised of Chapter 1 which introduces the problems that will be examined and explains the organization of the book. It also presents definitions of common terms and concepts that will be relied on throughout the text.

1 The Need for a Unified Conceptual Framework in Professional Psychology

Psychology has been remarkably successful as a scientific discipline since its founding in the late 1800s. Since World War II, the application of psychological science in clinical practice has been tremendously successful as well. Already in 1961, E. Lowell Kelly, the past president of the American Psychological Association (APA) Division of Clinical Psychology at that time, declared that the growth of clinical psychology was "well nigh phenomenal. Before World War II, clinical psychologists were few in number, poorly paid, and had but little status . . . . Ours is a success story without counterpart in the history of professions" (p. 9). Professional psychologists played a small role in health care before 1945 when the first licensure law for psychologists was enacted in the United States. Since then, the field has grown dramatically and now plays a major role in behavioral health care (e.g., there are now over 85,000 licensed psychologists in the United States; Duffy et al., 2006).

Despite the dramatic success of professional psychology, the field has also been marked by substantial controversy and conflict. There has been remarkable diversity in the theoretical orientations used to understand human psychology and the goals and processes of psychotherapy, and there has been deep conflict and competition between theoretical camps and schools of thought throughout the entire history of the field. Conflicts between schools and camps have subsided in recent years as a result of the development of integrative approaches to psychotherapy and other factors (these issues will all be discussed more fully in subsequent chapters). Nonetheless, the field is still characterized by wide diversity in the conceptualization of personality, psychopathology, and mental health treatment.

Explaining the nature of human psychology and the processes involved in psychotherapy and behavior change has proven to be a formidable challenge for behavioral scientists. Research has provided reliable explanations for many psychological processes, but other aspects of the tremendous complexity of human psychology have been difficult to unravel and are currently understood only in broad outline. This is particularly true for the more complicated processes that are often the focus of psychotherapy. Detailed descriptions of many basic psychological phenomena are widely accepted (e.g., with regard to sensation, perception, and the basic processes of cognition, affect, learning, and development), but there remains

Foundations of Professional Psychology. DOI: 10.1016/B978-0-12-385079-9.00001-1

a great deal to be learned about many highly complex processes such as the development of personality characteristics, the causes of psychopathology, the nature and assessment of intelligence and personality, and the mechanisms that account for psychotherapeutic change.

Professional psychology may be reaching a transition point, however. In recent years, research examining several aspects of psychological development and functioning has been progressing steadily. Major advances in areas ranging from genomics to sociocultural factors are improving our understanding of many important aspects of psychological development and functioning. Research has also verified the effectiveness of psychotherapeutic interventions at levels that compare favorably with those in medical, educational, and other types of human services.

A particularly important development is the improved validity of recent research findings. The results of recent research are not being challenged like they were in the past because the quality of research methodology has improved significantly. There is now broad acceptance of the need for stronger evidence to support the validity of inferences and conclusions, both in terms of research as well as clinical practice. The time is past when it was generally acceptable to argue for the superiority of one's theoretical or therapeutic approach only on the basis of one's past experience or data from uncontrolled research studies, rationales that naturally led quickly to disagreements and controversies. Standards regarding the validity of assessment findings, therapy outcomes, and research conclusions are all higher than they were even a decade ago.

Consensus has not yet emerged regarding several important areas of disagreement in the field, but there is no question that the tenor of recent professional disagreements is much different from those in the past. The 1990s, for example, were marked by highly contentious controversies involving repressed memories of child abuse, empirically validated treatments, the validity of quantitative versus qualitative research findings, and new treatments such as Eye Movement Desensitization and Reprocessing and psychotherapy for multiple personality disorder. The intensity of current disagreements pales in comparison.

Given these signs that professional psychology may be entering a new period in its development as a profession, it is important to revisit the basic frameworks that are used to conceptualize and organize education and practice in the field. The foundational conceptual frameworks that any profession uses, whether implicit or explicit, have a major impact on education, research, and practice. Each profession needs to ensure that those frameworks remain current with scientific and other developments, both within and outside the profession. Given recent developments in science, health care practice, and society generally, now is a good time for professional psychology to reexamine these issues.

There are several indications that the basic frameworks the field has traditionally used to conceptualize education and practice in professional psychology need to be updated. The next section describes several of the problems and sources of confusion that arise from the use of these frameworks. The educational setting within which graduate students begin their entry into the profession provides a useful context for illustrating the pervasiveness and significance of these problems.

Traditional Approaches to Professional Psychology Education and Practice

Students of psychology all learn about the wide variety of theoretical orientations that are used to explain psychological development, psychopathology, and psychotherapy. Very early in their coursework, students also learn that these orientations are typically based on assumptions or first principles that present quite different views on the nature of human psychology (e.g., fundamentally conflicted drives in the case of psychoanalysis, a blank slate in the case of behaviorism, an optimistic self-actualizing tendency in client-centered therapy, a postmodernist constructivism in solution oriented therapy). These widely varying perspectives on fundamentally important aspects of human nature naturally gave rise to disagreements regarding the validity of the differing approaches. Clearly, the phenomena under study in psychology are tremendously complex, so it would be expected that many different approaches to understanding those phenomena would be proposed. Nonetheless, a bewilderingly diverse array of over 400 different approaches to understanding personality, psychopathology, and/or psychotherapy has now been developed (Corsini & Wedding, 2008).

This complicated educational landscape is quite challenging for students to navigate. In addition to the many irreconcilable conflicts between the traditional theoretical orientations, new orientations continue to be developed. There has been significant growth in terms of integrative and eclectic approaches (Norcross, 2005), and completely new approaches are developed on a regular basis as well (e.g., acceptance and commitment therapy, positive psychotherapy, personality-guided relational psychotherapy; Hayes, Strosahl, & Wilson, 1999; Magnavita, 2005; Seligman, Rashid, & Parks, 2006). The shortcomings and weaknesses of the various approaches are commonly discussed in students' standard textbooks and controversies regarding the validity of these approaches are well known. Students also notice that few of the theoretical orientations incorporate scientific findings regarding the genetic underpinnings and biological functioning of the human mind and brain, and many do not fully integrate the evidence regarding the impact of sociocultural influences on behavior and development. And yet a comprehensive scientific approach to understanding human psychology would seem to require that all these factors be integrated into one's theoretical orientation for understanding human development, functioning, and behavior change. (These issues are discussed in detail in the chapters that follow.)

Despite the very complicated and confusing professional literature they are studying, students nonetheless are faced with a relatively urgent need to identify a theoretical orientation that they will use to organize and structure their approach to clinical practice. If their faculty does not inform them about which theoretical orientation would be the "correct" one to select, students are typically advised to choose one or perhaps more of the available orientations on their own. When students apply for practica and internships, for example, they are almost always asked about the theoretical orientation they use to guide their clinical practice. One of the required essay questions on the Association of Psychology Postdoctoral and Internship Centers (APPIC) *Application for Psychology Internship* asks them to "Please describe your theoretical orientation and how this influences your approach to case conceptualization and practice" (APPIC, 2009, p. 22). Students are normally quite practical about approaching this very important selection. Choosing an orientation that is not one of the most frequently endorsed approaches will almost certainly reduce one's chances of obtaining these required training placements and later employment positions. There are important reasons to choose one of the most popular orientations even if one judges it to be inferior in important respects.

How should students approach this defining feature of their emerging professional identity and career paths? Is it advisable to adopt just one approach in order to maximize the consistency and coherence of one's clinical work, or will one's practice be limited without competence in multiple orientations? Is it problematic to adopt an eclectic or common factors approach to practice? If the available orientations are limited in important respects, to what extent can one (or should one) modify an existing approach in order to make it more valid and useful, or would that be viewed as an inappropriate use of an established approach? In light of the recent movements toward evidence-based practice and competency-based accreditation and licensure, which theoretical orientation (or combination or modification of existing approaches) is the most likely to lead toward eventual success in the field? It can be risky to consult with graduate faculty and clinical supervisors about these questions because they naturally hold allegiances to their preferred theoretical orientations, and students depend on these individuals for grades, evaluations, and recommendations for later professional positions.

These issues raise challenging questions not only for students entering the field, but also for the strength and coherence of the theoretical and empirical foundations of the profession itself. It was perhaps inevitable that the rapid growth and development of the young discipline of psychology would lead to a variety of different theoretical perspectives on human psychology. These perspectives are not reconcilable with each other, however, and often receive only partial scientific support. If professional psychology is to be a science-based health care profession, current scientific findings need to be fully integrated into the frameworks and theoretical orientation that the field uses to conceptualize education and practice in professional psychology.

It is time for the profession to reexamine these issues. Indeed, one could argue that the need to address and resolve these issues has grown critical. The movements toward evidence-based practice and competency-based education, along with the growing scientific understanding of human psychology and the current economic outlook for health care, are quickly increasing pressures to resolve these issues. The lack of a common framework for conceptualizing professional practice in psychology introduces confusion and inefficiencies for students, accreditation and licensure bodies, professional organizations, universities, governmental bodies, insurers, as well as the general public. None of this benefits the profession or the public that we serve.

This volume attempts to show that the science and practice of psychology have progressed to the point where a unified approach to conceptualizing human development, functioning, and the practice of psychology is now possible. It argues that it is time to leave behind the era of conflicting theoretical approaches to understanding psychological practice and replace them with a unified science-based framework. To be sure, this would represent a major break from several important traditional practices and would be a significant transition for the profession. Nonetheless, recent developments compel the field to consider taking these steps.

Two Critical Questions Facing the Profession

There has not yet been extensive discussion within the field regarding a unified framework that psychologists from across the specializations and theoretical camps can apply as the foundational framework for conceptualizing professional psychology education and practice. To reach consensus on this issue, common ground from across the theoretical camps and practice areas will need to be identified. In addition, there are two fundamental questions that need to be considered. Consensus on a unified framework for the field is unlikely to develop without examining these two questions specifically.

One of these questions concerns the definition of the field. The above overview of confusing practices in professional psychology education suggests that the current definition of the field does not clearly identify the nature, scope, and purposes of professional practice in psychology. There is often disagreement regarding the appropriate approach to conceptualizing clinical cases and the type of treatment to implement. Answers to these types of questions depend heavily on who one asks. For example, if students enter certain graduate programs, they will learn that particular theoretical orientations are appropriate for conceptualizing clinical cases, but if they enter other programs they will learn that other theoretical orientations are appropriate for applying to the same types of cases. Other students enter programs where they will develop broad-based knowledge and skills and learn about a wide range of interventions, including even nonpsychotherapy interventions such as bibliotherapy, support group participation, meditation training, physical exercise, and the incorporation of religious, spiritual, or cultural practices. As a result, there is not consensus regarding what students should be competent to do as a result of completing their graduate training, which range of individuals they should be competent to work with, or what issues they should focus on in assessment and treatment. A clearer definition of the field may be necessary before consensus on these questions can develop.

The other critical question for the field concerns the scientific basis underlying professional psychology. A unified conceptual framework for the profession requires more than just clarity regarding the basic nature, scope, and purposes of psychological intervention. Such a framework also requires consensus regarding the scientific basis for clinical intervention. At first glance, it might seem completely unnecessary to raise questions about this issue—many would argue that the scientific basis of professional psychology is self-evident and obvious to anyone who practices the profession. Others, however, would raise questions regarding the strength of the scientific evidence justifying psychological intervention. In fact, there would be little argument that we possess only incomplete answers at this point regarding many important aspects of psychological development, functioning, and behavior change. At present, the causes and cures for many psychological syndromes are unclear, and many psychological processes are understood only in broad outline. A unified conceptual framework for the field may not be possible until there is greater clarity regarding the strength of the current scientific underpinnings of professional psychology and their implications for clinical practice.

These questions are discussed in detail in the following chapters. As will be shown, the answers to these questions have major and direct implications for education, practice, and research in the field. Consensus on these issues was perhaps not possible in the past given the state of the available knowledge regarding human psychology and behavioral health care. But the science and practice of psychology have evolved significantly in recent years, and the field may be reaching a transition point where it is now possible to unite around a common unified conceptual framework for professional psychology education and practice. Achieving such a unified perspective would be a momentous development for the field, filled with both difficult transitions and great potential. Resolving these questions and reaching a unified, science-based perspective may also be necessary, however, for the field to continue its remarkable trajectory of growth and development as a profession and a scholarly discipline.

Organization of This Volume

This book is divided into four sections that together cover the conceptual foundations of the field and their application in clinical practice. Part I ("Introduction") introduces the problems that will be addressed and the approach that will be taken, while Part II ("Conceptual and Theoretical Foundations of Professional Psychology") provides a more detailed analysis and evaluation of the theoretical and conceptual foundations of the field. Part III ("Conceptualizing Psychological Treatment from a Biopsychosocial Perspective") illustrates how a unified biopsychosocial framework can be applied clinically across the treatment process, and Part IV ("Additional Implications for Professional Psychology") addresses additional implications of this approach for professional psychology education and practice.

Chapter 2 argues that the current definitions of professional psychology used in the field introduce confusion about the nature and purposes of psychological intervention and that a more clearly specified definition will help resolve several issues that have long divided the field. Such a definition is then proposed. Chapter 3 illustrates how that definition would bring greater focus to professional psychology education and practice. Applying this definition would result in several significant changes in the way that the field approaches the assessment and treatment of behavioral health issues and biopsychosocial needs.

Chapter 4 examines the basic reasons for the wide diversity of theoretical orientations that have been developed in psychology and the theoretical and conceptual confusion that resulted as a consequence. This discussion explains the reasons for what has been the complicated and confusing *pre-paradigmatic* nature of the field. This is followed by a discussion of the basic requirements that a unified conceptual framework for the field would need to meet to provide a satisfactory theoretical approach for the profession. It is then argued that current scientific explanations for human psychology are now sufficiently detailed to support a unified, paradigmatic, science-based approach to understanding psychological development and behavioral health practice.

Chapter 5 discusses the scientific foundations of professional psychology from a modern biopsychosocial perspective, along with the major implications of this approach for education and practice in the field. As will be seen, the biopsychosocial approach is based on general systems theory, which, along with cybernetics and several other approaches, provided the conceptual basis for many scientific advances in recent decades. These approaches, taken together as a group, are known as nonlinear dynamical systems theory or complexity theory, and have become essential for understanding complex phenomena across the sciences and for understanding living systems in particular.

Chapter 6 emphasizes the critical role of ethics in professional psychology. Professional ethics must be incorporated into the foundational conceptual framework for the field because scientific knowledge alone cannot address many critical aspects of providing health care to the public. This chapter takes a theoretical foundations approach to professional ethics. In addition to emphasizing the importance of ethics codes, guidelines, and legal requirements, issues that are obviously critical in professional practice, this chapter argues that familiarity with the foundational principles and theories underlying these codes, guidelines, and legal requirements is necessary for analyzing and evaluating ethical issues comprehensively. The need for a deeper appreciation of ethical theory is growing as psychologists practice, conduct research, and teach in an increasingly diverse society and interconnected global community where technological, scientific, and social developments are presenting new challenges and opportunities.

Part II of the book ends with an integration of the issues discussed in Chapters 2–6. A unified framework for understanding the practice of psychology is presented in Chapter 7 that builds on and integrates the findings from the previous chapters. The definitional, scientific, and ethical issues discussed in the previous chapters provide the foundations for this integrated framework.

Part III discusses the application of the conceptual and theoretical framework presented in the earlier chapters to the psychological treatment process. The biopsychosocial approach has often been advocated as a superior approach to conceptualizing mental as well as physical health. There has been relatively little description, however, of how a biopsychosocial approach would be applied across the general as well as specialized areas of practice for the field as a whole. Therefore, this section of the book describes the psychological treatment process as approached from a biopsychosocial perspective and includes chapters on assessment, treatment planning, treatment, and outcomes assessment (i.e., Chapters 8–11, respectively).

Part IV begins with a chapter on prevention. It is often noted in professional psychology that the prevention of mental and physical disorders and the promotion of mental and physical health need to be higher priorities in American health care. This is especially evident when taking a biopsychosocial perspective to conceptualizing health and health care. These topics often receive limited emphasis in traditional professional psychology education, however. Following that discussion, the last chapter of the book revisits the potential of a unified conceptual framework for the field and its implications for education, licensure, practice, and research as the profession moves into the paradigmatic stage of its development.

Basic Definitions

The next chapter notes that professional psychology has often not used a standard vocabulary for communicating within the profession. Particular schools of thought have sometimes used terms with meanings and connotations that are not shared by other schools. This can make discussions regarding overarching theoretical issues that span the entire field quite difficult. To help avoid that problem here, the basic terminology used in this book is defined and the decisions to use certain terms over others are explained below.

Behavioral Health

Behavioral health normally has a broader meaning than *mental health*. While "mental health care" usually refers to the treatment of psychiatric disorders excluding substance dependence, "behavioral health care" normally includes substance abuse treatment and health psychology (also known as behavioral medicine, which involves using psychological services to address medical problems; see Blount et al., 2007). This volume focuses on behavioral health more broadly.

Biopsychosocial Approach

The biopsychosocial approach is a comprehensive, integrative framework for understanding human development, health, and functioning. It is based on the perspective that "humans are inherently biopsychosocial organisms in which the biological, psychological, and social dimensions are inextricably intertwined" (Melchert, 2007, p. 37). It is a science-based metatheoretical perspective that integrates a full range of psychological, biological, and sociocultural perspectives.

Engel (1977) offered the original formulation of the concept, which was based on general systems theory, a framework that has now been incorporated into modern complexity theory approaches to the scientific understanding of complex phenomena.

Complexity Theory

This is the commonly used term to refer to a broad class of complex mathematical modeling approaches that were developed over the past century to understand complex phenomena in the sciences. Also referred to as *nonlinear dynamical systems theory*, these approaches include general systems theory, cybernetics, chaos theory, network theory, fractals, self-organization and catastrophe theory, and other approaches to understanding complex adaptive systems. *Nonlinear dynamical systems theory* is the term often used to refer to this general category of approaches for less complex systems, while *complexity theory* is used to refer to more complex systems with more variables. Chapter 5 discusses these approaches in more detail.

Client Versus Patient

The people who psychologists serve when providing mental health services are usually referred to as clients or patients, and sometimes consumers or even customers. The term *client* is often used to refer to individuals who purchase products or services from businesses. The term has the advantage of avoiding the pathology focus of traditional health care where ill patients can be relatively passive recipients of the services provided by expert health care professionals. This traditional health care perspective can minimize the active role that psychotherapy clients generally need to take for treatment to be optimally effective. The term *patient* avoids the connotation of the business relationship where a desired or needed service is being purchased and where the ethical obligations the seller has are normally lower than in the case of health care. The term *patient* is also clearly associated with health care and conveys the sense of the special ethical obligations that health care professionals take on when providing services to their patients. Because of the health care orientation that is advocated in this volume, the term *patient* will be used throughout. It is emphasized, however, that the use of this term does not imply that the patient should passively accept the "sick role" or the expertise of health care professionals. Instead, the term is used to help emphasize the role and responsibilities of professionals who provide health care to the public.

Evidence-based Practice

This term refers to the application of empirical research findings in the health care treatment process. The APA Presidential Task Force on Evidence-Based Practice (2006) developed the following definition: "*Evidence-based practice in psychology* (EBPP) is the integration of the best available research with clinical expertise in the context of patient characteristics, culture, and preferences" (p. 273). This definition extends slightly the one developed by the Institute of Medicine in 2001 for application to all health care: "Evidence-based practice is the integration of best research evidence with clinical expertise and patient values" (p. 147).

General Versus Specialized Practice

General practice psychology tends not to be specifically defined in professional psychology or the other mental health fields. Because of the health care focus of the book, however, differences between general and specialized practice are important to the discussion. *General practice* will be used here to refer to services offered to meet the relatively common mental health needs of the public in general. The term usually refers to working with young and middle-aged adults, though some might include children and/or seniors in their conceptualizations of general practice. (In both medicine and behavioral health care, including children and seniors in general practice is more common in rural areas and underserved urban areas where fewer specialists are available.) *Specialized practice*, on the other hand, includes those areas that require more detailed knowledge and skills regarding a narrower range of issues and/or populations. Some specializations have more formal definitions and training requirements (e.g., school or neuropsychology), whereas others are less well defined (e.g., couples therapy, forensic psychology).

Mental Health

Mental health generally refers to the psychological functioning of individuals. The mental health fields have often focused on mental disorders, problems, and deficits, but healthy and optimal functioning are also critical in the full spectrum of mental health. The term *mental health care* usually does not include substance abuse treatment or health psychology—areas that are normally included when using the term *behavioral health care*. This volume focuses on behavioral health more generally and consequently refers to mental health less frequently.

Nonlinear Dynamical Systems Theory

Please see the description of *complexity theory* above. The term *nonlinear dynamical systems theory* is often used in discussions of less complex phenomena. It is also used more commonly by scientists, while *complexity theory* is more commonly used by the wider public for referring to the same general category of scientific approaches.

Professional Psychology

Professional psychology involves the clinical application of psychological science and professional ethics to address behavioral health needs and promote biopsychosocial functioning. (See Chapter 2 for a detailed discussion of this definition.)

The emphasis of this volume is on the field of professional psychology as a whole, which in the United States normally refers to clinical, counseling, and school psychology as well as all the specialized subfields subsumed within these broader areas of clinical practice. Unfortunately, the field of psychology does not use a single-word term that differentiates the *science* of psychology from the *practice* of psychology similar to the way, for example, that biology is differentiated from medicine or physics from engineering. Despite the wordiness, *psychological science* will generally be differentiated from *professional psychology* throughout the book.

Psychological Intervention

This term refers to all the direct behavioral health care services offered by professional psychologists, including the full range of psychological assessments, psychotherapies, psychoeducational and other psychosocial interventions, and outcomes assessments that are used in behavioral health care. These interventions are employed across a wide variety of settings ranging from hospitals to outpatient clinics, independent practices, schools and colleges, military installations, rehabilitation institutes, and nursing homes to address problems and disorders and improve health and functioning across the biopsychosocial domains (e.g., medical health and physical functioning, behavioral health, family functioning, and educational and vocational effectiveness).

Therapist

The behavioral health field includes several specializations (primarily clinical, counseling, and school psychology, psychiatry, mental health counseling, clinical social work, marriage and family therapy, and rehabilitation counseling). Most of the foundational issues discussed in this volume apply across these behavioral health care specializations in general, though the discussion focuses on issues particularly relevant to professional psychology. Because of the foundational perspective taken in this volume and the broad applicability of the issues discussed, reference will often be made to psychotherapists in general, as opposed to referring only to professional psychologists. *Therapist* will also often be used as the short form for *psychotherapist*.

Part II

Conceptual and Theoretical Foundations of Professional Psychology

This part of the book addresses the two issues identified in the last chapter as critical to resolving the theoretical and conceptual confusion that exists in the profession. It evaluates the conceptual and theoretical foundations of professional psychology in terms of their historical development and their current adequacy in light of contemporary scientific knowledge and health care practices. Chapters 2 and 3 focus on the basic definitions of professional psychology that the field has relied on to help identify the nature and scope of practice in the profession. It then shows how a clearer definition that emphasizes the health care purposes of psychological intervention would clarify many aspects of professional psychology education and practice. Chapters 4 and 5 examine the development of the theoretical foundations of the field, and emphasize how those foundations need to evolve to be consistent with the current scientific understanding of human development and functioning. Chapter 6 focuses on the ethical foundations of behavioral health care, because no conceptualization of health care is complete that does not incorporate an ethical perspective. These various perspectives are then integrated in Chapter 7 to present a unified conceptual framework for education, practice, and research in professional psychology.

2 Professional Psychology as a Health Care Profession

This chapter addresses the first of the two key issues that were identified in the previous chapter as needing resolution before consensus can develop regarding a unified framework that will resolve the theoretical and conceptual conflicts and confusion that have long divided professional psychology. This issue concerns the lack of a clear definition of the field. As explained below, professional psychology developed in a highly diversified manner that allowed great latitude in the ways that personality, psychopathology, and treatment could be conceptualized. A wide range of theoretical orientations was developed to guide the assessment and treatment of psychological issues and problems, and there has long been heated controversy surrounding the ways that disorders and problems are conceptualized, the appropriateness of particular interventions, and the best way to train professional psychologists. These disagreements were often quite divisive and resulted in major fractures within the field.

There are multiple reasons why professional psychology developed in this complicated manner. Chapter 4 will focus on reasons for the complicated theoretical and scientific development of the field as it attempted to understand the very complex subject matter of psychology, while this chapter focuses on the definitions the field has used to conceptualize the nature and scope of clinical practice in psychology. After proposing a revised definition for the field, Chapter 3 will then illustrate how this new definition could be applied to better inform clinical practice and education. The examination of these issues begins with a brief historical review of past approaches to defining the field in order to clarify the nature and significance of the problems that need to be addressed.

Traditional Approaches to Defining Professional Psychology

One of the most notable features of the field of professional psychology has been the diverse array of theoretical orientations that have been used to understand its subject matter. As noted in Chapter 1, hundreds of different theoretical orientations for understanding personality, psychopathology, and psychotherapy have been developed. Many of these orientations fundamentally conflict with each other, and their limitations as explanations for psychological phenomena are widely known. This

Foundations of Professional Psychology. DOI: 10.1016/B978-0-12-385079-9.00002-3

has also resulted in psychologists using very different terms and concepts for understanding personality, psychopathology, and the goals and processes of psychotherapy (e.g., the terminology and concepts used by psychodynamic, behavioral, humanistic, and postmodern theoretical orientations often vary widely). As a profession, the field has developed in a highly decentralized manner, like a loose federation of camps that share a common general purpose but diverge regarding many of the particulars. These camps have generally been united in terms of using psychological principles and primarily verbal means to achieve behavioral, cognitive, and/or affective changes (biofeedback and psychopharmacology represent exceptions), but have otherwise used highly varied approaches to understanding personality, psychopathology, and treatment.

There have been advantages and disadvantages associated with this highly diversified approach to understanding the field. One advantage was that the field avoided prematurely committing to a single theoretical approach that was later found to be inadequate. The field of psychiatry could be viewed as having taken this path when it first strongly embraced psychoanalytic theory, and then a second time when psychopharmacology replaced psychoanalysis as the dominant approach. Another advantage of the diverse approaches taken in professional psychology is that they led to great creativity and ingenuity in the application of psychological interventions to a wide variety of problems and issues. Recent examples extending outside behavioral health care include executive coaching and sports psychology.

There have been disadvantages associated with this highly diversified approach to practicing psychology as well. These were already evident in the early stages of the development of the profession, and their impact is still being felt today. Because of their importance to the present discussion, these disadvantages will be outlined in more detail.

Following World War II, when professional psychology began getting established as a profession, the 1949 Boulder Conference was convened in response to the rapid growth of professional psychology in the Veterans Administration. The recruitment of large numbers of psychologists into this new health care system was a major impetus for the growth and development of the young profession (Benjamin, 2007). An important purpose of the conference was to identify a common core of training for clinical psychologists who would be eligible to work in that system. Although the conferees agreed that there should be a common core for professional psychology education, they also argued that there was not "one best way" (Raimy, 1950, p. 55) to train clinical psychologists and strongly suggested that this issue be left to university programs to decide. The primary point of agreement at the conference was the view that learning to conduct research was necessary for developing competence as a practitioner, and the "scientist—practitioner" model quickly became the dominant approach throughout clinical and counseling psychology. Disagreements grew regarding the appropriateness of this model, however, and eventually became very serious, leading to the breaking away of the National Council of Schools of Professional Psychology (NCSPP) in 1974. This organization took a fundamentally different approach, emphasizing clinical competencies while de-emphasizing research skills (Peterson et al., 1992). There is some overlap between the scientist—practitioner and professional models of training, but there is also a major divergence in the basic conceptualization regarding how psychologists should be trained to practice the profession.

While these early formulations of the nature of professional psychology were being developed, a wide variety of theoretical approaches to practice were also being developed. These included variations of psychodynamic, humanistic, behavioral, cognitive, and postmodern orientations that soon became incorporated into psychologists' clinical practices and students' clinical training. This led to highly multifaceted but confusing practice and educational environments. For example, different students within the same educational program might take fundamentally different approaches to case conceptualization and intervention with patients, sometimes resulting in completely different assessment findings, treatment plans, and courses of treatment. Despite the irreconcilable theoretical differences between many of the approaches taken, students and clinicians were able to find theoretical and/or empirical support to justify the use of each of them.

In 1975, in the very influential case where a licensure applicant sued regarding his denial to sit for the psychology licensure examination, Judge MacKinnon of the District of Columbia Court of Appeals ruled that the tremendous latitude allowed in the conceptualization of professional psychology education and practice was not defensible. He found that professional psychology was an "amorphous, inexact, and even mysterious discipline [and] possession of a graduate degree in psychology does not signify the absorption of a corpus of knowledge as does a medical, engineering or law degree" (as cited in Wellner, 1978, p. 6). The implications of this decision were serious. Consequently, the National Register of Health Service Providers in Psychology and other psychology organizations collaborated on the development of the core curriculum requirements, which then became incorporated into American Psychological Association (APA) accreditation and most state licensing board standards. These include the familiar domains still reflected in current APA Commission on Accreditation (CoA) requirements, namely professional ethics and standards; research design; statistics; measurement; history and systems; individual, biological, cognitive, affective, and social bases of behavior; and completion of an internship (Wellner, 1978).

These general core curricular domains brought some much-needed consistency to professional psychology education. Nonetheless, the previous emphasis on allowing programs substantial latitude to decide their particular training models and curricular goals and objectives persisted under the new guidelines. In fact, this practice continues to the present day. The current APA Commission on Accreditation's [CoA] (2009) *Guidelines and Principles for Accreditation of Programs in Professional Psychology* explicitly state that:

> *The accreditation guidelines and principles are specifically intended to allow a program broad latitude in defining its philosophy or model of training and to determine its training principles, goals, objectives, desired outcomes (i.e., its "mission"), and the methods to be consistent with these. Stated differently, the CoA recognizes that there is no one "correct" philosophy, model, or method of doctoral training for professional psychology practice; rather there are multiple valid ones. (p. 4)*

Furthermore, although the CoA guidelines do state that "Broad and general preparation for practice at the entry level" (p. 3) is required for a program to be accredited, "broad and general preparation" is not further defined or delineated.

This practice of allowing great latitude in the goals and objectives of professional psychology education programs is reflected in other practices within the profession as well. For example, the current Association of Psychology Postdoctoral and Internship Centers (APPIC) *Application for Psychology Internship,* which is used by almost all internship programs listed by APPIC, includes the following among its required essay questions: "Please describe your theoretical orientation and how this influences your approach to case conceptualization and intervention" (APPIC, 2009, p. 22). Allowing students to choose the theoretical orientation they use to conceptualize their approach to professional practice is also reflected in the 2007 report of the APA Assessment of Competency Benchmarks Work Group. For example, the Work Group identified the following as an "essential component" for demonstrating intervention skills: "ability to formulate and conceptualize cases and plan interventions utilizing at least one consistent theoretical orientation" (p. 43). This approach was also taken by the National Council of Schools of Professional Psychology in its report on competencies for professional psychology education, also published in 2007, which identifies the following outcome for the development of intervention skills: "Understanding of the mutual influence of chosen theory and intervention on the process of therapy" (National Council of Schools of Professional Psychology, 2007, p. 17).

Allowing professional psychology education programs and individual practitioners great latitude in their chosen theoretical orientations may have been necessary to accommodate the diverse array of orientations that were used in the field. This decision, however, also led to a number of confusing practices within the profession. As discussed in Chapter 1, the practice of requiring professional psychology students to choose a theoretical orientation to guide their personal clinical practice, despite the fundamental theoretical contradictions between these orientations and their inconsistent empirical support, presents a complicated decision for many students. For example, in cases where one's academic advisor and the program director diverge in their theoretical orientations, is it safest to pick the orientation endorsed by one's advisor or the program director? Some students resolve this dilemma by referring to the advisor's preferred orientation when in advising meetings, an instructor's theoretical orientation when in class, and the clinical supervisor's orientation when at practicum or internship, and then employing their actual preferences following graduation.

Consider also the curricula used in professional psychology education. Competency-based approaches to education have become standard in psychology as well as other professions in recent decades (Nichols & Nichols, 2001). But how do programs implement a competency-based curriculum when students can decide on an individual basis the theoretical orientation they adopt, the particular competencies they subsequently develop, and the disorders and populations with which they will work using that particular theoretical orientation? The competencies needed to implement the various theoretical orientations can vary substantially.

While there is overlap among the knowledge and skills required to conduct psychodynamic, behavioral, cognitive, motivational interviewing, or dialectical behavior
therapy, there is also significant divergence. Furthermore, some theoretical orientations are recommended for use with particular populations or mental health issues.
As a result, do students or licensure applicants need to demonstrate competence
with any particular populations or diagnostic categories, or can they demonstrate
competence with just the populations for whom their theoretical orientation is most
appropriate? Should a student or licensure applicant demonstrate at least basic proficiency with the assessment of common mental health issues with some reasonably
diverse set of patients, even though he or she might later provide treatment to only
a particular group consistent with his or her preferred theoretical orientation?

There have been some attempts in recent years to develop consensus regarding
the types of competencies that professional psychologists should develop. The
report of the APA Assessment of Competency Benchmarks Work Group (2007)
describes outcomes from the most important of the recent efforts. It addressed six
categories of functional competencies that were viewed as necessary for the practice of psychology. The first three of these categories are clearly critical for mental health practice (assessment/diagnosis/case conceptualization, intervention, and
consultation). The relevance of the last three categories, however, is less clear
because they involve skills less closely related to clinical practice. These three are
research/evaluation, supervision/teaching, and management/administration, and
include "essential components" such as "generation of knowledge ... engages in
systematic efforts to increase the knowledge base of psychology through implementing research" (p. 50), "evaluation of effectiveness of learning/teaching strategies
addressing key skill sets" (p. 58), and "demonstrate leadership skills and abilities,
business knowledge, management and supervisory skills needed to develop system"
(p. 60).

These latter three competencies are obviously critical for psychologists working
as researchers, teachers, and administrators, but including non clinical skills as
essential core competencies raises questions about the scope of training and practice that is appropriate, practical, or necessary for professional psychologists.
Students rarely take courses in teaching or administration, for example. Students or
interns who are competent in clinical practice but not necessarily in research, teaching, and administration are routinely allowed to graduate, and state psychology
boards do not disqualify licensure applicants who cannot demonstrate proficiency
in each of these areas. The members of the APA Task Force on the Assessment of
Competence in Professional Psychology (Kaslow et al., 2007) emphasized the difficulty of making further progress with a competency-based approach to professional
psychology education when they concluded that "Probably the most challenging
and yet most foundational recommendation [for implementing such an approach] is
that the profession develop a consensus regarding the definition of the *core competencies*" (p. 448; see also Cummings & O'Donohue, 2008).

Resolving these issues is becoming more and more important. The lack of a
clear definition of the profession leads to confusion regarding the knowledge and
skills that should be taught in professional psychology education, accreditation

standards for professional psychology education programs, licensure requirements, and the mission and goals for professional organizations in the field. This then contributes to confusion among governments, institutions, other health care providers, and insurance companies regarding the role of professional psychology in health care. It also fails to provide clear guidance to individual practitioners for how to approach their clinical practices. The next section addresses these questions in more detail before a definition of professional psychology is proposed that is intended to resolve these issues.

Defining Professional Psychology

The seminal Boulder scientist–practitioner model for clinical psychology specifically avoided any delineation of a core curriculum or practice competencies for the field. During the time of the Boulder Conference, professional psychologists were employing a growing variety of competing theoretical orientations in their clinical practices. This situation naturally made it difficult to identify specifically what constituted appropriate mental health care practice. As long as the services provided were psychological in nature, these various approaches were considered to fall within the purview of professional psychology. This is reflected in the broad definition of professional psychology that is still used by the APA CoA, which states that for purposes of accreditation "'professional psychology' is defined as that part of the discipline in which an individual, with appropriate education and training, provides psychological services to the general public" (APA CoA, 2009, p. 2). The emphasis here is on providing services that are psychological in nature, as opposed to, for example, providing health care services that meet the behavioral health needs of the general public. The latter would define the field as a health care profession, while the former merely requires that members of the profession provide services that are psychological in nature without any specification of the goals or purposes of those services.

During this same time when professional psychology was formally getting established, the field was beginning to be formally recognized as a health care profession that serves the mental health needs of the general public. The passing of legislation requiring that psychologists become licensed had a particularly important impact in this regard. In 1945, Connecticut became the first state to license psychologists, and by 1977 all 50 states required licensure. Across the United States, the profession is now state-sanctioned and regulated to provide safe and competently delivered behavioral health care to the general public. Public and private health care systems have come to rely heavily on government-controlled and -administered licensure and regulatory procedures.

The evolving identity of professional psychology as a health care profession has been recognized relatively slowly. The mission statement of the APA for its first half-century of existence (1892–1945) referred only to the scientific purposes of psychology. In 1945, the statement was expanded so that the purpose of the organization was "to advance psychology as a science [and] as a profession" (Wolfe, 1946, p. 721). Another half-century later, in 2001, the word "health" was added to affirm the status of professional psychology as a health care profession (Johnson, 2001). It then asserted that the mission of the APA "shall be to advance psychology as a science and profession and as a means of promoting health and human welfare." The 2009 APA Presidential Task Force on the Future of Psychology Practice Final Report further noted that "The Task Force supports this trend [toward psychology as a health care profession] and notes it is time to make a clear commitment to our identity as a health care profession as differentiated from solely a mental health profession" (APA Presidential Task Force, 2009, p. 3).

Though psychology's role as a health care specialization has not been clear within the profession, there does appear to be general agreement among governmental bodies and professional organizations that the basic purpose of professional psychology is to address the behavioral health needs of the general public. The following definitions of the profession are currently in use by the APA CoA, the National Council of Schools of Professional Psychology (NCSPP), the Association of State and Provincial Psychology Boards (ASPPB), the APA Commission for the Recognition of Specialties and Proficiencies in Professional Psychology (CRSPPP), and the American Board of Professional Psychology (ABPP).

- APA CoA Guidelines and Principles (2009, p. 2): "For the purposes of this document, 'professional psychology' is defined as that part of the discipline in which an individual, with appropriate education and training, provides psychological services to the general public."
- NCSPP Core Curriculum Conference Resolutions (Peterson et al., 1992, p. 159): "The primary goal of education for professional psychology is preparation for the delivery of human services in a manner that is effective and responsive to individual needs, societal needs, and diversity."
- ASPPB Model Act (1998) for the licensure of psychologists: "A health service provider in psychology means an individual licensed under this Act who is duly trained and experienced ... in the delivery of direct preventive, diagnostic, assessment, and therapeutic intervention services to individuals whose growth, adjustment, or functioning is actually impaired or is demonstrably at high risk of impairment."
- APA CRSPPP (2011a) definition of clinical psychology: "Clinical Psychology is a general practice and health service provider specialty in professional psychology."
- APA CRSPPP (2011b) definition of counseling psychology: "Counseling psychology is a general practice and health service provider specialty in professional psychology."
- ABPP (2011a) specialty certification in clinical psychology: "Clinical psychology is both a general practice and a health service provider specialty in professional psychology. Clinical psychologists provide professional services ... [to] ... individuals across the life-span."
- ABPP (2011b) specialty certification in counseling psychology: "A Counseling Psychologist facilitates personal and interpersonal functioning across the life span ... using preventative, developmental, and remedial approaches, and in the assessment, diagnosis, and treatment of psychopathology."

Taken together, these official definitions indicate significant consensus around the view of professional psychology as a health care field that serves the general public. Although the emphases of the above definitions differ, all but two of them specifically focus on the role of professional psychologists as being health service providers. This is certainly the view of state licensing boards, governmental health care programs, and health insurance companies that reimburse and regulate psychologists almost exclusively in their role as health care providers.

As professional psychology became more widely recognized as a health care field, it also began to develop specialized areas of practice. As research findings accumulated and psychological principles were applied in larger numbers of settings, various professional organizations developed definitions and guidelines for areas such as school, child, and medical psychology, neuropsychology, and, more recently, psychopharmacology. These specializations are relatively clear about the knowledge, skills, and training experiences that are needed for practice within their respective areas, often specifying a full range of graduate, internship, and postdoctoral training requirements. In contrast, the general practice areas have continued to allow more latitude in the theoretical orientations, the treatment approaches, and the disorders and populations with which one can develop clinical competence.

Drawing clear lines between general and specialized areas of professional psychology practice would require substantial analysis and is not a main purpose of this discussion. In general, however, expectations for understanding a broad range of clinical issues and demographic populations are greater for those in general practice, while those in the specializations are expected to possess significantly greater knowledge regarding a narrower range of issues and populations. In either case, however, a basic understanding of the mental health needs of the public in general is necessary. Even a therapist who works exclusively with children, for example, also needs to work with the children's parents and integrate an understanding of adult psychopathology, family functioning, and parenting in order to appropriately assess and treat children's issues. Likewise, one who works with seniors needs to be able to work effectively with their children (for those who have them) and possess an understanding of lifespan development in order to intervene effectively. Broad, foundational knowledge of individual and family development, psychopathology, and biopsychosocial functioning across the lifespan is consequently necessary for working with individuals of all ages and circumstances.

Given the confusion and problems associated with the lack of clarity surrounding the nature, scope, and purposes of the profession, a definition of professional psychology is proposed below that more clearly specifies these factors. This definition integrates components of the official definitions noted earlier while emphasizing the perspective of professional psychology as a health care profession that aims to meet the behavioral health needs of the general public. To be clear, this definition begins by emphasizing the applied nature of professional psychology as a profession that is based on psychological science and professional ethics. It then goes on to describe the fundamental purpose of professional psychology when viewed as a health care specialization. It also incorporates a biopsychosocial emphasis, the importance of which will become clearer in the next three chapters. To be thorough, it also notes that the field is composed of general and specialized areas of practice.

A Proposed Definition of Professional Psychology

Professional psychology is a field of science and clinical practice that involves the clinical application of scientific knowledge regarding human psychology and professional ethics to address behavioral health needs and promote biopsychosocial functioning. As a health care specialization, it provides psychological services to meet the behavioral health and biopsychosocial needs of the general public. It includes general as well as specialized areas of practice.

Discussion

One might argue that meeting the mental health needs of the general public is the obvious role for professional psychology—why is it even important to point this out? This definition, however, is quite different from the APA CoA definition that emphasizes providing psychological services without any specification of the purpose of those services. Specifying the purpose of the services (i.e., meeting the behavioral health needs of the public) has several critical implications. If the profession is defined as simply providing psychological services to the public, then a professional psychologist can learn one of the available theoretical orientations and offer services based on that orientation to individuals who request them. From this perspective, the profession is defined primarily as a service industry where patients need to take responsibility for their choices regarding whether to purchase particular services. For example, an individual concerned about some depressive symptoms can seek out a variety of treatments, such as cognitive behavioral therapy, online positive psychology interventions, antidepressant medication, or over-the-counter St. John's wort. From the perspective of a service industry, the legal principle of *caveat emptor,* or "let the buyer beware," applies—it is buyers who bear the primary responsibility for making their purchases of products and services and for evaluating whether they meet their needs.

Taking a health care perspective on the field implies a significantly different approach, however. Providing health care to meet the needs of the general public carries obligations to provide care that is safe and effective. Instead of purchasing services that individuals have decided to try, patients are now entrusting mental health care professionals to accurately diagnose their behavioral health needs within the context of their biopsychosocial circumstances and to develop treatment plans and provide interventions that will address their needs and promote their welfare and well-being (see also Chapter 6). From this perspective, the focus is on the interventions that will be helpful for meeting the needs of individual patients. Therapists' personal preferences regarding theoretical orientations, for example, do not play a large role when professional psychology is defined in this manner. Instead, the underlying science and ethics of behavioral health care play the major role.

The implications of the above definition for professional psychology are substantial, and so they will be described in detail in the following chapters.

The scientific and ethical foundations of professional psychology specified in this definition are discussed in Chapters 4–6, while the implications of focusing on the mental health and biopsychosocial needs of the general public are discussed in the next chapter. These considerations are then integrated into a unified conceptual framework in Chapter 7.

Applying this definition in the specific context of the behavioral health needs and biopsychosocial circumstances of the general public will further clarify the implications of this definition for education and practice in professional psychology. That is the subject of the next chapter.

3 The Public We Serve: Their Mental Health Needs and Sociocultural and Medical Circumstances

The primary purpose of this chapter is to illustrate how the definition of professional psychology proposed in the previous chapter can clarify the nature, scope, and purposes of professional psychology education and practice. Greater clarity regarding these issues will be very helpful for resolving several of the disagreements regarding the appropriate approaches to teaching and practice that have divided the field. Before different camps and schools of thought will be able to come together around a unified theoretical framework for professional psychology, there needs to be more agreement regarding the basic purposes, scope, and underlying rationale for clinical practice in the field.

The definition of professional psychology proposed in the previous chapter identified the primary function of a professional psychologist as providing psychological services to meet the behavioral health and biopsychosocial needs of the general public. When focusing on behavioral health needs, it might seem to be a straightforward process to learn the assessments and interventions that are most effective for treating common behavioral health problems. Epidemiological data could be used to identify the mental health disorders most commonly faced by the public, and then the available research could be reviewed to identify the assessments and interventions that are most effective for addressing those issues. Treatment guidelines could then be written that would suggest the indicated treatments for various mental health disorders.

As all seasoned therapists know, however, this seemingly straightforward approach is complicated by the many biopsychosocial considerations that need to be taken into account when treating behavioral health problems. For example, there is probably agreement that a general practice psychologist should be able to assess and treat major depression in adults because it is one of the most common mental health problems that individuals encounter. But psychologists must also be familiar with a wide variety of additional issues in order to appropriately diagnose and treat major depression, including co-occurring substance abuse, other Axis I conditions, Axis II disorders, physical health and medical conditions, stressors, a variety of sociocultural factors, and level of functioning in important life roles. All these factors can have a major impact on assessment, treatment planning, the course of treatment, and of course also the outcomes of treatment.

Foundations of Professional Psychology. DOI: 10.1016/B978-0-12-385079-9.00003-5

To illustrate how the definition of professional psychology proposed in the previous chapter clarifies the complex nature of clinical practice in psychology, this chapter provides a comprehensive biopsychosocial perspective on meeting the behavioral health needs of the general public. It illustrates how a biopsychosocial approach clarifies the nature and range of behavioral health and biopsychosocial issues that fall within the purview of professional psychology. This perspective is critical to the discussion of a theoretical framework that can unify the field around a common approach for understanding human psychology and behavioral health practice. The discussion begins with a consideration of the mental health needs of the general population, followed by a review of the broader sociocultural and medical circumstances that also need to be incorporated into this perspective.

Behavioral Health Needs

Thoroughly reviewing the behavioral health needs of the general population would require substantial analysis. Even just a brief overview of the issues, however, provides an informative perspective on the behavioral health problems and concerns currently faced by the general US population. A good place to start is with the available epidemiological data.

It is clear that behavioral health problems are common among the public in general. In the largest study of comorbidity ever conducted in the United States, the National Comorbidity Survey found that nearly 50% of the respondents reported at least one lifetime mental disorder and nearly 30% reported at least one 12-month disorder (Kessler et al., 1994). The most common disorders were depression, alcohol dependence, social phobia, and simple phobia. Further, a subgroup of 14% of the respondents accounted for the vast majority of the severe disorders and had three or more comorbid disorders over the course of their lifetimes. This study was replicated a decade later, and the National Comorbidity Survey Replication found that the prevalence of mental disorders had changed little from the earlier study (Kessler et al., 2005; Wang, Demler, Olfson, Wells, & Kessler, 2006; Wang et al., 2005). Among its findings were the following:

- 50% of all Americans report having symptoms diagnostic of a mental disorder during their lifetime.
- Many of their symptoms emerged as early as age 11; half of all lifetime cases started by age 14 and three quarters started by age 24.
- More than one-fourth of adults reported having symptoms diagnostic of a mental disorder over the previous year, and most of these disorders could be classified as at least moderate in severity.
- Mental illness is the most prevalent chronic health condition experienced by youth.
- Most people wait years or even decades to seek treatment for their depression, anxiety, or bipolar disorder.
- Fewer than one-third of those with mental disorders receive adequate treatment for their mental health problems.

The very high prevalence of mental disorders, their seriousness, and the low frequency of treatment suggest that much greater emphasis should be placed on providing behavioral health care, care should be provided when individuals are young and first experiencing mental health problems, many more individuals should be brought into treatment, and greater emphasis should be placed on the prevention of behavioral health problems. These implications will be addressed at various points later in the book.

To provide a more detailed illustration of the behavioral health issues faced by the general population in the United States, the prevalence data reported in the *Diagnostic and Statistical Manual of Mental Disorders* (4th ed., text revision; DSM-IV-TR; American Psychiatric Association, 2000a) for diagnoses with 1% or greater prevalence are summarized in Table 3.1. The cutoff of 1% is arbitrary but helpful for considering whether a problem might be viewed as relatively common and within the purview of general as opposed to specialized psychology practice.

Table 3.1 Psychiatric Disorders Rank-Ordered by Prevalence as Reported in the DSM-IV-TR

Disorder	Prevalence
Hypoactive sexual desire	33% females[a]
Premature ejaculation	27%[a]
Nicotine dependence	Up to 25%
Female orgasmic disorder	25%[a]
Alcohol dependence	15%
Major depressive disorder	10−25% women, 5−12% men
Specific phobia	7.2−11.3%
Social phobia	3−13%
Male erectile disorder	10%[a]
Male orgasmic disorder	10%[a]
Adjustment disorder	Up to 50% of those who experience a specific stressor
Acute stress disorder	14−33% of people exposed to severe trauma
Body dysmorphic disorder	5−40% of individuals with anxiety or depressive disorders
Learning disorders	2−10%
Primary insomnia	1−10%
Breathing-related sleep disorder	1−10%
Posttraumatic stress disorder	8%
Attention-deficit/hyperactivity	3−7% school-age children; data limited for adults
Dysthymic disorder	6%
Generalized anxiety disorder	5%
Cannabis use disorders	Almost 5%
Hypochondriasis	1−5%
Circadian rhythm sleep disorder	0.1−4% adults; up to 60% of night-shift workers

(Continued)

Table 3.1 (Continued)

Disorder	Prevalence
Pathological gambling	0.4−3.4% adults; 2.8−8% adolescents and college students
Schizotypal personality disorder	3%
Antisocial personality disorder	3% males, 1% females
Bulimia nervosa	1−3% of females
Histrionic personality disorder	2−3%
Dementia	1-year prospective prevalence of 3% in adults
Obsessive-compulsive disorder	2.5%
Paranoid personality disorder	0.5−2.5%
Cocaine use disorders	2%
Borderline personality disorder	2%
Panic disorder	1−2%
Somatization	0.2−2% of women
Bipolar I	0.4−1.6%
Amphetamine use disorders	1.5%
Schizophrenia	0.5−1.5%
Obsessive-compulsive personality disorder	1%
Mental retardation	1%
Dependent personality disorder	Among the most frequent personality disorders seen in mental health clinics

[a]Reported are prevalence estimates of sexual complaints. The DSM-IV-TR notes that it is unclear whether these complaints meet diagnostic criteria for a sexual disorder.

These data suggest that the most prevalent DSM-IV-TR disorders and complaints experienced by the general population involve sexual functioning and substance abuse. These findings are themselves perhaps not surprising, but it is remarkable how relatively little attention these subjects tend to receive in professional psychology coursework and clinical training. In order of decreasing prevalence, after several depressive and anxiety-related disorders are body dysmorphic disorder, learning disorders, and sleep disorders, all topics that tend to receive limited coverage in professional psychology education.

Particularly when viewed from a holistic, biopsychosocial orientation to behavioral health care, these prevalence data have clear implications for professional psychology education and practice. More attention needs to be given to a full range of behavioral health problems if psychologists are going to be well prepared to address the behavioral health needs of the general public. These questions are still more complicated, however, because examining behavioral health only in terms of the presence of psychiatric disorders provides a seriously limited perspective. Additional perspectives are needed to gain a thorough understanding of behavioral health and biopsychosocial functioning.

Suicidality is an important public health issue that is not clearly reflected in the above prevalence data. Suicide was the 11th leading cause of death in the United States in 2004, accounting for 1.4% of all deaths (Center for Disease Control, 2007). It accounted for 12.7% of all deaths among 15−24-year-olds and was the second leading cause of death among 25−34-year-olds. It is estimated that up to 40% of the general population in the United States have had suicidal ideation at some point in their lives (Chiles & Strosahl, 2005; Hirschfeld & Russell, 1997; Zimmerman et al., 1995), and one large survey of college students found that slightly more than half reported having had suicidal ideation at some point (Drum, Brownson, Denmark, & Smith, 2009). It has also been estimated that perhaps 10−12% of adults report having made at least one suicide attempt (Chiles & Strosahl, 2005).

Another perspective necessary for understanding the behavioral health needs of the general population involves positive mental health. Keyes (2007) has emphasized that the absence of mental illness does not imply the presence of mental health. In fact, there are several indications that anything less than a state of positive mental health (which Keyes labels "flourishing") is associated with increased functional impairment and increased physical and behavioral health problems. Based on national survey data, Keyes estimated that roughly only 2 in 10 Americans are flourishing, while nearly 2 in 10 are in poor mental health, which he refers to as "languishing." Most of the rest are in between. That nearly 8 in 10 Americans are not experiencing positive mental health represents a major public health issue. The majority of these individuals also do not have a psychiatric disorder and consequently are not likely to come to the attention of behavioral health care professionals and receive treatment.

In summary, by starting with a definition of professional psychology that focuses on meeting the behavioral health and biopsychosocial needs of the general public, the data summarized above point to a significantly broader range of behavioral health needs than what is typically addressed in professional psychology education and clinical training. Professional psychology education currently tends not to focus extensively on several of the most prevalent behavioral health issues, such as problems related to sexual function, substance dependence, lack of positive mental health, and even suicidality. Issues related to chronicity and comorbidity are also highly prevalent and obviously have major impacts on individuals' functioning and treatment. These often receive insufficient attention in graduate education as well, however.

The above overview clearly suggests that the public faces a broader and more complicated range of behavioral health problems than what is addressed in many traditional professional psychology education programs. In addition, however, there are several additional factors that need to be integrated into the discussion. Individuals' behavioral health concerns cannot be understood in isolation—their broader life context must also be considered in order to gain a sufficiently complete understanding of their development and current functioning and the behavior change interventions that will be effective for addressing their problems and promoting their well-being.

Sociocultural and Medical Circumstances and Characteristics

There is widespread agreement across health fields at this point that a full range of interacting psychological, sociocultural, and biological factors need to be considered to gain a comprehensive understanding of individuals' psychological development and functioning. Omitting important factors from any of these domains can result in inaccurate diagnoses and assessments as well as ineffective treatment plans. (See the two chapters that follow for a detailed discussion of these issues.)

For purposes of illustration, below is an overview of several of the factors that need to be integrated into case conceptualizations in professional psychology when approached from a biopsychosocial perspective. As with the case of behavioral health concerns, many of these data are generally well known. Nonetheless, a brief review can be informative when considering the nature and scope of professional psychology education and practice. The categories below were chosen because they are included in the Joint Commission on Accreditation of Healthcare Organizations [JCAHO] (2006) standards for behavioral health care facilities, the five-axial DSM-IV-TR system (American Psychiatric Association, 2000a), as well as other basic approaches described in standard textbooks for learning psychological assessment and intervention.

Demographic Characteristics

When professional psychology is viewed as a health care profession serving the general public, basic census data highlight the necessity of being able to work with a variety of demographic groups. In the 2000 US Census (US Bureau of the Census, 2000), 25.7% of the population were under age 18 and 12.4% were age 65 or older; 12.7% were military veterans; and 75.1% reported that their race was White, 12.5% Latino, 12.3% African American, 3.6% Asian, and 0.9% Native American. English only was reported as spoken at home by 82.1% of the population, and 87.7% reported that they were born in the United States.

Based on current trends, the US Bureau of the Census projects that racial and ethnic minority individuals will make up the majority of the US population by 2042 and minorities will represent half of all children by 2023 (US Bureau of the Census, 2008). The US population is becoming increasingly diverse, and knowledge and skills for dealing with the cultural influences and challenges faced by individuals who are not members of the mainstream culture need to be incorporated into the competencies required for professional psychology practice. Failing to do so will result in the profession becoming irrelevant to increasing numbers of individuals who do not fall within the traditional target groups for many traditional psychotherapeutic treatments (Sue & Sue, 2008).

Medical Conditions

Professional psychologists are not responsible for assessing and treating medical conditions, but they are responsible for generally understanding the interaction of physical health and psychological functioning as well as knowing when referrals for medical evaluations may be needed. Consequently, obtaining information regarding patients' medical health and functioning is incorporated into standard approaches to psychological assessment.

A brief survey of prevalence data finds that distressing and disabling medical conditions are remarkably common in the US population. In fact, nearly 50% of the US population live with a chronic health condition that requires routine adherence to prescribed treatment regimens and/or involves activity limitations (Partnership for Solutions, 2004). Table 3.2 shows the prevalence of specific medical conditions experienced by more than 1% of adults (Center for Disease Control,

Table 3.2 Medical Conditions in US Adults Rank-Ordered by Prevalence (Those Greater Than 1%), as Reported in the CDC (2005) National Health Interview Survey

Disorder	Prevalence (%)
Overweight	35
Lower back pain	28
Chronic joint symptoms	27
Obesity	25
Hypertension	22
Arthritis, rheumatoid arthritis, gout, lupus, or fibromyalgia	21
Restlessness	18
Hearing difficulty without hearing aid	17
Nervousness	16
Limitations in physical functioning	15
Neck pain	15
Migraine/severe headaches	15
Felt everything was an effort	14
Sinusitis	13
Heart disease	12
Sadness	11
Asthma	11 (7 still have it)
Vision trouble (even with correction)	9
Hay fever	9
Absence of all natural teeth	8
Diabetes	7
Coronary heart disease	7
Cancer	7
Ulcer	7
Hopelessness	6
Worthlessness	5
Face/jaw pain	4
Chronic bronchitis	4
Stroke	2
Emphysema	2
Kidney disease	2 (in past 12 months)
Underweight	2
Liver disease	1 (in past 12 months)

2005). Many of these problems tend to cause pain and/or limitations in one's ability to carry out social roles and responsibilities, and virtually all of them have major psychological components in terms of etiology, treatment, and/or consequences. These problems consequently have major implications for behavior and functioning across the biopsychosocial domains. These data also further highlight the importance of taking a biopsychosocial approach to behavioral health care.

Behavior and lifestyle factors have become the leading causes of morbidity and mortality in the United States. Smoking is the leading cause of preventable morbidity and mortality and accounts for 18.1% of all deaths, while obesity has become the second leading cause of preventable morbidity and mortality and accounts for 16.6% of deaths (Mokdad, Marks, Stroup, & Gerberding, 2004). Alcohol consumption is the next most important factor (3.5% of deaths). Loeppke et al. (2009) further identified the leading health conditions that contribute to decreased economic productivity and increased health care costs in the United States—the five leading conditions were depression, obesity, arthritis, neck/back pain, and anxiety. In addition to the very large amount of treatment that is needed to address all these cases, these data strongly argue for a preventive orientation to health care, with much greater attention given to behavior and lifestyle.

The importance of a biopsychosocial perspective for understanding physical health is increasingly being recognized by the medical establishment in the United States. For example, the Institute of Medicine (2004) concluded that roughly 50% of morbidity and mortality in the United States is caused by behavior and lifestyle factors, and that medical school curricula need to increase the coverage of behavioral influences on physical health if physicians are going to increase their effectiveness at treating medical problems. As medicine increases its attention to the influence of behavior on physical health, so should professional psychology, and more attention should also be given to the influence of physical health on behavior and biopsychosocial functioning.

Educational Attainment, Vocational and Financial Status

Problems associated with educational attainment, employment, and financial resources have important impacts on mental health and biopsychosocial functioning. The prevalence of these types of problems is also high. The census data presented in Table 3.3 show that one in five Americans have not attained a high school diploma or equivalency diploma. In addition, one in seven Americans lack basic English literacy skills such as those needed to read a map, review a paycheck for accuracy, or understand a warning label on a tool or medicine bottle (National Center for Educational Statistics, 2008).

Large numbers of families also experience significant financial stress. Surveys find that financial difficulties and stress are the leading cause of marital problems in the United States (Vyse, 2008). Extremely serious problems with financial insecurity are prevalent as well; for example, the lifetime prevalence of homelessness in the United States is estimated to be 6.2% (Toro et al., 2007). Professional

Table 3.3 Selected Educational and Vocational Results from the 2000 US Census

Educational Attainment	
Less than 9th grade	7.5%
9th−12th grade, no diploma	12.1%
High school graduate or GED	28.6%
Some college, no degree	21.0%
Associate degree	6.3%
Bachelor's degree	15.5%
Graduate or prof. degree	8.9%
Income	
< $10,000	9.5%
$10,000−14,999	6.3%
$15,000−24,999	12.8%
$25,000−49,999	29.3%
$50,000−99,999	29.7%
$100,000−200,999	9.9%
$200,000 and over	2.4%
Median family income	$50,046
Living in Poverty	
Families with children	13.6%
Single females with children	34.3%
Single females with children <5	46.4%

psychology education typically gives minimal attention to these important influences on psychological and family functioning.

Family Characteristics and Relationships

Family, relationship, and parenting problems are common and can have critical effects on individual and family development and functioning. Divorce affects roughly half of the ever-married population. Of couples in their first marriage, approximately 33% divorce or separate within 10 years, and the rate of divorce is higher for subsequent marriages. Over half of divorces occur in families with minor children, affecting over 1 million children each year. Cohabiting couple relationships are even more likely to dissolve than marriages, and 46% of these include minor children (Blaisure & Geasler, 2006). In addition, nearly 25% of American women report being raped and/or physically assaulted by a former or current spouse, cohabiting partner, or date at some time in their lifetime (Centers for Disease Control and Prevention and National Institute of Justice, 2000).

In addition to the many signs of marital and relationship difficulties experienced by couples and families, there are many signs of difficulties with parenting as well, and these have major implications for psychological development and behavioral health. The next section briefly notes the prevalence of child maltreatment. Another very influential (and related) factor is attachment. Research finds that one's attachment style is associated with a wide range of outcomes across the psychological, interpersonal, and social realms including emotional regulation, behavioral self-regulation, psychopathology, interpersonal and relationship functioning, sexual functioning, behavioral functioning in work and organizational settings, and engagement in psychotherapy (Cassidy & Shaver, 2008; Mikulincer & Shaver, 2007). One's own attachment during childhood also appears to have a strong effect on later parenting as an adult. Parents' perceptions of their own childhood attachments have been found to predict the attachment classification of their children 75% of the time (Main, Kaplan, & Cassidy, 1985; Steele, Steele, & Fonagy, 1996). Research finds that attachment style is also relatively enduring. The continuity of attachment from infancy to adulthood is moderately stable, while the continuity of attachment across adulthood is quite stable (e.g., on average, approximately 70% of adults received the same attachment classification across time periods extending up to 25 years; Mikulincer & Shaver, 2007). Given the stability and impact of attachment style, it is disheartening that only about 65% of infants in the general population are found to be securely attached, with the remaining 35% distributed among the insecure classifications (i.e., avoidant, anxious-ambivalent, and disorganized; Prior & Glaser, 2006). All these issues need to be incorporated into the knowledge base that psychologists bring to professional practice.

History of Child Maltreatment

The US Department of Health and Human Services, Administration on Children, Youth and Families (2008) reported that approximately 6.0 million children were reported to child protective service agencies in 2006 for alleged maltreatment. This represents approximately 8% of all American children. Approximately 62% of these reports were investigated, and 30% of these resulted in a determination that at least one child had been the victim of abuse or neglect. With regard to child physical abuse, a nationally representative study of youth aged 10–16 years found that 22% reported experiencing a non family assault and 7.5% reported a family assault in their lifetime (Finkelhor & Dziuba-Leatherman, 1994). After reviewing the available data on sexual abuse, Finkelhor (1994) estimated that at least 20–25% of women and between 5% and 15% of men experienced child sexual abuse. After reviewing the available data on emotional and psychological abuse, Binggeli, Hart, and Brassard (2001) estimated that more than one-third of the US adult population may have experienced psychological maltreatment, and between 10% and 15% experienced the more severe and chronic form of this type of maltreatment.

There is extensive evidence regarding the effects of child maltreatment on psychological development and functioning, and rates of child maltreatment are routinely found to be much higher in clinical samples than they are in the general population (Myers et al., 2002). Although behavioral health practitioners are generally well aware of the significance of child maltreatment in patients' lives, professional psychology education often does not cover this topic extensively.

Legal and Criminal Involvement

Many types of criminal victimization and legal involvements are highly stressful and traumatic, and adjustment and acute stress disorders occur frequently among those experiencing specific stressors and trauma (see Table 3.1). The US Department of Justice National Crime Victimization Survey (US Bureau of Justice Statistics, 2007) found that 2.1% of individuals age 12 and over experienced a violent crime in 2005 (which was a dramatic drop from the 1970s when rates were routinely greater than 4.5%). An estimated 15.4% of households also reported a property crime in 2005 (which was a dramatic drop from the 1970s when rates were consistently above 50%). In addition, 7.2 million individuals were in prison or jail or on probation or parole in 2006, and nearly 200,000 tort suits were filed in 2003 in the 15 states reporting these data. Crime and legal involvement affects a significant proportion of the population and consequently is incorporated into standard approaches to psychological evaluation. This topic also tends to receive little emphasis in professional psychology education, however.

Religion and Spirituality

Surveys consistently find high levels of religiosity in the United States compared with other Western countries. Approximately 76% of the adults interviewed in the American Religious Identification Survey identified as Christian (Kosmin & Keyser, 2009), while the Baylor Religion Survey found that 89% of the respondents were affiliated with a congregation, denomination, or other religious group (Bader et al., 2006). A Newsweek/Beliefnet (2005) poll found that 55% of the respondents reported being religious and spiritual, another 9% reported being religious but not spiritual, and another 24% reported being spiritual but not religious. A fourth major poll conducted by Financial Times/Harris found that 73% of the American adults surveyed believed in God or some other type of supreme being (Harris Interactive, 2006). Despite the importance of this factor in the lives of many individuals, this is another area that often receives limited emphasis in professional psychology education.

Implications for Professional Psychology as a Health Care Profession

Each of the biopsychosocial factors noted above can have a major impact on one's development and current functioning. These factors can all have a significant influence on the course of psychological treatment as well. As a result, these topics have been integrated into the commonly accepted approaches to assessment and treatment planning used in the behavioral health field (e.g., the five-axis DSM-IV-TR diagnostic system, the JCAHO standards for the accreditation of behavioral health care facilities). A large number of the topics reviewed above do not receive extensive coverage in many professional psychology education programs, however. In addition, the treatment of co-occurring disorders and the ongoing treatment and management of chronic behavioral health and medical conditions are also typically addressed in a limited manner in professional psychology education. From the biopsychosocial perspective on health care emphasized in this volume, all these issues need to be integrated into the knowledge and skills that professional psychologists bring to clinical practice. Individuals' development and functioning and the behavior change process cannot be thoroughly understood without taking this type of integrative, holistic perspective.

Employing a definition of professional psychology as a biopsychosocially oriented health care profession would focus educational programming on the broad range of behavioral health needs and biopsychosocial circumstances faced by the general public. Epidemiological data such as those reviewed above could be used to focus the curriculum on the specific knowledge and skills needed to meet those needs. Professional psychology students do acquire knowledge and clinical experience dealing with all these issues because they cannot be avoided in clinical training—these are the problems that patients bring with them into treatment. The acquisition of too much of this knowledge and skill occurs in a haphazard manner, however, and often outside formal coursework and clinical training. Taking a biopsychosocial health care perspective on the profession would result in the systematic integration of these topics into professional psychology education and practice.

4 Understanding and Resolving Theoretical Confusion in Professional Psychology

Psychology as a science and a profession developed very rapidly over the past century, but not without going through major growing pains. On the one hand, the achievements of psychology have been very impressive. Its impacts on the social sciences, mental health care, and even society and culture generally have been remarkable—few academic disciplines have gained such a far-reaching influence in such a short period of time. On the other hand, there have been deep divides between schools of thought and theoretical camps regarding the validity of competing theoretical orientations for understanding personality, psychopathology, and the goals and processes of psychotherapy. The competition and conflicts between these schools and camps have often been intense and have resulted in major fractures and no unified voice for advocating for the field. The existence of so many different theoretical orientations and schools of thought within the field is not as odd as it might first appear when viewed from a historical perspective (as will be seen below). Nonetheless, the range and variety of conflicting approaches present a complicated and confusing picture for those inside as well as outside the profession.

To understand how psychology developed in such a complicated and confusing manner, this chapter examines the development of scientific fields in general. The complicated development of psychology as a discipline is generally much more understandable when viewed from the perspective of the development of the natural and social sciences. Several of the attempts to resolve theoretical confusion within psychology are then discussed, followed by a discussion of the basic requirements that a comprehensive solution to this problem would need to meet. Before addressing these questions, the nature of the problem of theoretical confusion in the field will first be outlined in more detail.

The Complicated Theoretical Setting Within Professional Psychology

The many theoretical orientations that have been formulated to explain human development, functioning, and behavior change are well known within the

Foundations of Professional Psychology. DOI: 10.1016/B978-0-12-385079-9.00004-7

profession. Many of the textbooks that review these orientations take a chronological approach, starting with Freudian theory and progressing through psychodynamic, behavioral, humanistic, feminist, multicultural, and other approaches. These orientations vary in the thoroughness of their explanations regarding personality, psychopathology, and psychotherapy. Some theories, for example, focus primarily on the development of personality and psychopathology, with little emphasis on the process of psychotherapy (e.g., existentialism), while others focus primarily on therapy processes and methods (e.g., solution-oriented and narrative therapies). The theory providing the most comprehensive and thorough explanations for personality, psychopathology, and psychotherapy has been Freudian psychoanalysis, which also ended up being the most controversial of the theories.

In addition to the individual unitary theories, a variety of eclectic and integrative approaches have been developed. These, too, take diverging perspectives. Norcross (2005) has categorized these into technical eclecticism (where theory is relatively unimportant), theoretical integration (where two or more of the traditional theories are combined), common factors (where therapeutic factors common to all therapies are emphasized), and assimilative integration (where techniques from a variety of orientations can be assimilated into one of the traditional theories). These eclectic and integrative approaches have contributed a great deal to psychotherapy research and practice, but they also offer additional competing options regarding the appropriate conceptualization of personality, psychopathology, and treatment.

The proliferation of theoretical approaches in professional psychology is also continuing. New eclectic and integrative approaches continue to be developed (Norcross, 2005) as well as entirely new approaches such as positive psychotherapy (Seligman, Rashid, & Parks, 2006), acceptance and commitment therapy (Hayes, Strosahl, & Wilson, 1999), attachment therapy (Wallin, 2007), and personality-guided relational psychotherapy (Magnavita, 2005). Indeed, over 400 different theoretical orientations have now been developed (Corsini & Wedding, 2008). This very large number and the continuing proliferation of approaches raise challenging questions about the state of theory in professional psychology. As Beutler (1983) noted, newer orientations are often developed specifically because the earlier ones are viewed as inadequate. In addition, no individual theoretical approach has become dominant, not even an eclectic or integrative one. Surveys normally find that the largest number of adherents to any one orientation, even an eclectic or integrative approach, still remains a minority, usually less than one-third of the sample (Norcross, 2005).

The validity of these various theoretical orientations has been hotly debated over the entire history of the field. These orientations are often based on foundational assumptions or philosophical first principles that take widely varying perspectives on human nature (e.g., biologically based drives and conflicts in Freudian theory, the blank slate of nearly complete malleability in behaviorism, an optimistic self-actualizing tendency in humanistic theories, a postmodern constructivism in solution-focused therapy, Buddhist principles in the "third wave" of behavioral therapies emphasizing mindfulness). These philosophical starting points often conflict in fundamental ways and lead to very different conceptualizations of the nature of human development and psychological treatment. Naturally, heated disagreements then arise regarding the validity of the different theoretical orientations, and the divides between theoretical camps and schools of thought have often been very wide. Driver-Linn (2003, p. 270) noted that "Perceptions of psychology as beleaguered by fractionation and uncertainty are almost ubiquitous." In fact, many leading psychologists have been concerned that the ongoing conflicts and fragmentation of the field are so serious that its continued viability as a scholarly discipline may be threatened (e.g., Gardner, 2005; Kendler, 2002; Rychlak, 2005; Staats, 2005; Sternberg, 2005).

This is obviously a discomforting situation for the field. On the one hand, there is widespread consensus that many of the treatments based on these various theories are remarkably effective in clinical practice. The classic meta-analysis by Smith and Glass in 1977 found that all the tested therapies were effective, and virtually every meta-analysis conducted since then has reached the same conclusion (e.g., Grissom, 1996; Lambert & Bergin, 1994; Lipsey & Wilson, 1993; Wampold, 2001). On the other hand, the success of these theoretical orientations for explaining human development and functioning is limited. Many of the philosophical first principles or basic assumptions underlying these orientations have not been tested empirically and, when they have, the evidence supporting their validity is often mixed. The standard textbooks that review these orientations often explain that many are relevant for only particular populations or disorders, do not take into account important sociocultural factors, and/or have received only partial or inconsistent support as explanations for personality, psychopathology, and behavior change. Many theoretical orientations have not been systematically researched at all.

This complicated state of the literature in professional psychology leads to many confusing situations for students and practitioners. The field offers a remarkably diverse array of theoretical orientations for conceptualizing mental health, but each appears to explain only part of the whole story. There appears to be no consensus about the appropriate approach to take, and no combination or integration of approaches has yet been widely accepted. At the same time, demands that one's approach to clinical practice be supported by empirical evidence are increasing. This situation also makes it very difficult for the field to identify a set of common core knowledge and skills that should be taught in professional psychology education and could then provide the basis for program accreditation and psychology licensure (Kaslow et al., 2007; McHugh & Barlow, 2010).

The Evolution of Psychology and the Natural Sciences

Given this situation, one might conclude that it is simply premature to suggest that a unified theoretical framework for guiding professional psychology education and practice is possible. At present, the field lacks consensus regarding the appropriate theoretical orientation or conceptual framework for professional practice, and the various disagreements noted above continue to be widely debated. Before considering whether a unified theoretical orientation for the field can be developed, it is important to appreciate how the current state of affairs came to be. In light of how scientific fields develop in general, the variety of theoretical orientations that have been proposed in psychology is actually not nearly as unusual as it might first appear. In fact, in some ways it reflects the usual development of scientific fields.

Complexity of Psychological Phenomena

Before the complicated development of psychology as a discipline can be properly understood, it is critical first to appreciate the tremendous complexity of the subject matter involved. Without that understanding, the complicated and conflictual history of the field could be easily misinterpreted as a sign of serious scholarly weakness.

The observation that it is easier to explain phenomena in the so-called "hard sciences" (i.e., the physical sciences) than in the "soft sciences" (i.e., the social sciences; von Foerster, 1972) has become well known. Indeed, Staats (1999) noted that psychological phenomena are vastly more complex than the phenomena typically investigated in the natural sciences. The tremendous intricacy, complexity, and intangibility of phenomena in the social and behavioral sciences make human psychology a fascinating field of study and clinical practice, though empirically explaining these phenomena is highly challenging. Indeed, the human mind appears to be the most complex phenomenon that human beings have ever attempted to understand. Biologists have long emphasized the complexity of the human organism, declaring it the most complex system known to exist in the universe. Dawkins (1976, p. xxii), for example, concluded that "we animals are the most complicated and perfectly designed pieces of machinery in the known universe," while Wilson (1998, p. 81) stated that "the most complex systems known to exist in the universe are biological, and by far the most complex of all biological phenomena is the human mind." The proliferation of diverse theoretical orientations for understanding psychological phenomena and the lack of a unified conceptual framework within the field are much more understandable when the extraordinary complexity of psychological phenomena is taken into account.

Keeping this complexity in mind, an examination of the evolution of psychology as a scientific discipline reveals a progression that has been common across scientific fields. The two most influential approaches to understanding the development and evolution of scientific fields in general are (a) Thomas Kuhn's model of "scientific revolutions" and (b) the examination of the development of the conceptual and technological tools available to researchers over time. Of these two, the Kuhnian perspective has played a larger role in how psychologists have viewed the development of their discipline, and it will be reviewed first.

Kuhnian Scientific Revolutions

Early in their history, the natural sciences were characterized by an array of diverging and competing frameworks for understanding natural phenomena, and the movement toward consensus explanations of phenomena was often highly contentious. For example, Kuhn (1962) noted that during the first half of the 18th century "there were almost as many views about the nature of electricity as there were important electrical experimenters ... all were components of real scientific theories ... Yet though all the experiments were electrical and though most of the experimenters read each other's works, their theories had no more than a family resemblance" (pp. 13–14). Many scientific giants, such as Galileo, Newton, Cavendish, Watt, Lavoisier, and others, battled vehemently over the superiority and ownership of theories and explanations. Kuhn noted that the fragmentation and competition between researchers and theoretical camps when the physical sciences were young were very similar to that of the social sciences more recently.

Kuhn (1962) argued that the evolution of scientific fields tends to follow a five-stage pattern. The first stage is characterized by many conflicting views and competing explanations about what is even the proper focus of research within a field. This first stage is termed *pre-paradigmatic* because what is lacking is a paradigm, a major scientific achievement that convincingly explains phenomena in a particular area and around which the scientific community can unite. The first scientific field to leave this "immature" stage was physics, followed by chemistry and more recently biology. Kuhn argued that the second stage, which he called *normal science*, was reached when one school of thought or paradigm was found to explain phenomena better than the others. At this point, a great deal of time, sometimes decades or even centuries, is spent testing deductions made within the paradigm, with minor improvements being made as a result. If anomalies that are discovered can be accounted for by the existing theories, the paradigm is strengthened. If not, however, a third stage is reached. This *crisis* stage involves a period of "pronounced professional insecurity" (pp. 67–68) because there is no acceptable explanation that can account for the anomalies. During the fourth stage (*revolutionary science*), an active struggle ensues between the defenders of the old paradigm and proponents of a new paradigm, with each camp attempting to solve the greatest number of anomalies with their theories. In the final stage (*resolution*), one paradigm becomes dominant, which then generates a new period of *normal science* (stage two).

As a young field investigating tremendously complex phenomena, many areas within psychology have been pre-paradigmatic. Especially within professional psychology, numerous theoretical camps have been devoted to pursuing particular explanations of phenomena in an insular, parochial manner. Even when referring to similar or perhaps the same constructs, these camps have used different assumptions, terminology, and definitions of constructs (e.g., ego, ego-strength, self, self-concept, self-image, self-esteem, self-worth, self-confidence, self-efficacy). What Kuhn (1962) noted about the early research on electricity in the 18th century could also apply to psychology in the 20th century: though all the theories were examining the same phenomena, they "had no more than a family resemblance" (p. 14). And like the physical sciences early on, competition between theoretical camps in psychology has often been intense and contentious—Larson (1980) characterized this as the "dogma eat dogma" nature of the field.

The Kuhnian model of scientific revolutions is often cited as an explanation for the complicated development of psychology as an academic discipline (e.g., Driver-Linn, 2003). And there is no doubt that much of psychology has been pre-paradigmatic (Kuhn, 1962). The question of whether the field is ready to leave behind its pre-paradigmatic stage of development is central to the thesis of this book and will be explored later in this chapter. Before moving on to that question, a second view on the evolution of scientific fields provides another valuable perspective for understanding the evolution of psychology. This perspective has been quite influential in the natural sciences and takes a very different approach from Kuhn, instead emphasizing the tools that have been available to conduct scientific research.

Availability of Conceptual and Technological Tools

The ability of researchers to understand and explain phenomena is highly dependent on the power and precision of the tools that are available to them. From this perspective, scientific revolutions occur when new conceptual tools (e.g., calculus, statistics) or technological tools (e.g., telescopes, microscopes) are developed that allow for a more complete understanding of phenomena (Crump, 2001; Mitchell, 2009; Stewart, 1995). Over recent centuries, there have been truly remarkable advances in the capabilities of these tools.

The most important conceptual scientific tool ever discovered was mathematics, and the discovery of increasingly complex mathematics has resulted in profound scientific advances (Crump, 2001; Mitchell, 2009; Stewart, 1995). To illustrate, mathematics in Europe was written in words up until the invention of algebra. In 1202, Leonardo of Pisa published *Liber Abaci*, which enabled a shift from written to symbolic mathematics. The new numerical system allowed human "computers" to perform calculations that dramatically transformed the science and commerce of the time. The new system was so successful, seeming to capture the very essence of natural phenomena, that Galileo remarked that mathematics is the very language of nature. Newton later used algebra as the foundation for developing calculus, which was necessary to understand rates of change in motion and phenomena such as gravity. Calculus proved to be so useful that "The physicists went off looking for other laws of nature that could explain natural phenomena in terms of rates of change. They found them by the bucketful—heat, sound, light, fluid dynamics, elasticity, electricity, and magnetism" (Stewart, 1995, p. 16). Einstein viewed Newton's development of calculus as "perhaps the greatest advance in thought that a single individual was ever privileged to make" (as quoted in Capra, 1975, p. 56). Further advances in mathematics, and particularly the development of highly complex mathematical modeling over the past several decades, has allowed major advances in the understanding of complex adaptive systems (i.e., nonlinear dynamical systems and complexity theory; see the next chapter).

The invention of new technological tools (as opposed to conceptual tools) similarly transformed other areas of science. In 1543, Copernicus published his hypothesis that the earth revolved around the sun rather than the other way around, but it was not until Galileo built his first telescope in 1609 that this hypothesis could be confirmed. Galileo's first telescope had 8 × magnification, but within a few months he achieved 20 × magnification. This allowed him to discover, in just a matter of months, that the Milky Way consists of thousands of stars, the known planets are nearby, planets reflect sunlight, and several other phenomena (Crump, 2001). Further technological innovations in the telescope, including the Hubble Space Telescope, are continuing to transform our understanding of the cosmos (Dar, 2006).

The microscope has proven to be one of the most versatile scientific instruments ever invented (Crump, 2001). Van Leeuwenhoek (1632−1723) made the best early instruments, capable of 270 × magnification, and greatly expanded the boundaries of the observable world as a result. Van Leeuwenhoek subsequently discovered protozoa, bacteria, blood corpuscles, capillaries, and the circulation of blood through capillaries. He was the first person to observe spermatozoa and then found them in the males of all species that reproduce sexually. Biology was transformed as a result. The next revolutionary advance in microscopes occurred in the 1930s and used beams of electrons instead of beams of light. This allowed very small objects such as viruses, chromosomes, and nucleic acids (including DNA) to be observed, and biology was transformed once again.

Recent advances in science would be impossible without the electronic computer. The recent major advances in genetics, brain imaging, and particle physics, for example, require massive amounts of data processing in addition to highly complex technological equipment—the tremendous amount of data generated by the equipment would be useless without extremely high computing capacity. Computing capabilities have been increasing at an exponential rate ever since the integrated circuit ("chip") was invented in 1960. For example, the Sanger Center at Cambridge currently hosts 150 terabytes (150 trillion bytes) of unique genomic data and has processing power of about 2.5 teraflops (2.5 trillion operations per second). The Large Hadron Collider in CERN, Switzerland, has begun generating several petabytes (a thousand trillion bytes) of particle physics data each year. These types of research projects will actually generate more scientific data by several orders of magnitude than what has been collected in *all* of prior human history (Hey & Trefethen, 2003). Such capabilities are even transforming the way science is conducted in these areas. Instead of the usual approach, which involves "Hypothesize, design and run experiment, analyze result," the new approach involves "Hypothesize, look up answer in data base" (Lesk, 2004, p. 1).

Improvements in neuroscience research tools involving instrumentation, measurement, statistical and mathematical modeling, and infomatics are continuing to revolutionize science. For example, the very high spatial and temporal resolutions in magnetoencephalography brain scanning are making it possible to make highly precise measurements of neuronal activity many times per second (Cohen, 2004). This is in contrast to the one- or few-at-a-time measurements at low resolution that have been possible with functional magnetic resonance imaging machines. This difference is analogous to watching brain activity with a high-resolution video camera as compared to poorly focused still photos. This type of technological advance is revolutionizing brain research. Instead of investigating the "bottom-up" connections from one neuron to the next, or investigating "top-down" models such as the organization of intelligence or personality through the factor analysis of IQ or personality test data, it is becoming possible to investigate comprehensive, detailed, multilevel models that simultaneously combine bottom-up and top-down approaches in one model (Wood et al., 2006).

The increasing power and complexity of scientific tools, both technological and conceptual, are propelling revolutionary advances across many scientific fields. Many of the recent advances focus on the mind and brain in particular and have critical implications for the science and practice of psychology. Before examining those implications, it is important to note that understanding the reasons for the sometimes convoluted and lengthy development of the sciences in general shows that the complicated development of psychology follows a pattern that has been repeated in other disciplines and is also a logical consequence of the extraordinary complexity of the subject matter under investigation. Though the complicated development of psychology has often been difficult and frustrating for those inside the field, it has also been quite natural given these considerations.

The question at the center of the thesis of this volume is whether the field is ready to leave behind its complicated pre-paradigmatic stage of development and move on with a unified paradigmatic theoretical framework. Before addressing that question, it is important to be aware of the several past proposals that were developed to accomplish this goal. This question has been on the minds of psychologists for quite some time.

Clarifying Conceptual Confusion in Psychology

Psychologists have long been concerned about the irreconcilable differences between many of the theoretical orientations in the field and the problems this caused for clinical practice. Over the years, many solutions for resolving these differences have been offered. These proposals generally fall into two categories. One group has focused primarily on unifying approaches to the science of psychology while the other has focused primarily on integrating theoretical approaches to clinical practice.

One of the most ambitious proposals for a unified approach to the science of psychology was offered by Staats (1963, 1983), who connected findings from across the major subfields within the discipline. Anchored in evolutionary psychology, his model explained how humans develop basic behavioral response patterns that are in turn integrated into higher cognitive processes such as language. A variety of systemic theorists also emphasized interactions across levels of influence on individuals' development. The most influential of these was Bronfenbrenner's (1979) ecological framework, which incorporates a full range of individual, family, community, and cultural factors that are important in children's development. More recently, Sternberg and Grigorenko (2001) defined "unified psychology" as a multiparadigmatic and multidisciplinary study of psychological phenomena where findings are integrated from across disciplines and schools of thought. Henrique (2003) also proposed the Tree of Knowledge System, which focuses on the position that psychology holds among the natural and social sciences, where physics has focused on the nature of physical matter, biology on living organisms, psychology (specifically "psychological formalism") on the mental life of animals, and the social sciences on the nature of humans as self-conscious, social, and cultural beings. For recent discussions of these and other proposals for unifying the science of psychology, see Henrique (2005), Henrique and Cobb (2004), Henrique and Sternberg (2004), Magnavita (2008), Pinker (2002), Rubin et al. (2007), and Sternberg (2005).

The other general category of unifying proposals has focused primarily on clinical practice. There were some early attempts at identifying linkages between the major theoretical orientations used for practicing psychotherapy (French, 1933; Rosenzweig, 1936), and some very influential integrative approaches followed several years later (e.g., Dollard & Miller, 1950; Frank, 1961). A significant number of these approaches have now been developed, and Norcross (2005) has categorized them into technical eclecticism, theoretical integration, common factors, and assimilative integration according to how the processes of psychological development and/or behavior change are conceptualized. In addition, a variety of family systems theories have taken an integrative approach to understanding individual and family functioning (e.g., see Nichols, 1998). Magnavita (2005) also recently presented a major new unified approach to psychotherapy called personality-guided relational psychotherapy, which is based on *personality systematics*, the study of complex systems applied to personality functioning. For recent discussions of integrative and unified approaches that focus on the practice of psychology, see Anchin (2008), Henrique and Sternberg (2004), Magnavita (2006, 2008), and Norcross and Goldfried (2005).

These proposals have generated substantial research, provided important insights into the nature of human psychology and psychotherapeutic intervention, and provided useful alternative frameworks for clinicians in many types of practices. Analyzing the strengths, weaknesses, and potential of these various approaches for unifying the field extends beyond the scope of the present volume. (A great deal of this analysis has already been done—for more information, see the references noted above.) For present purposes, it will suffice to note that none of these approaches has been widely endorsed as a theoretical orientation that provides a comprehensive, integrated framework for guiding clinical practice in psychology. Even though several of these proposals have become widely known, none has been widely adopted as a framework for organizing and structuring education or practice in professional psychology. It appears that none of these proposals is widely viewed as satisfactory for resolving the conceptual confusion that pervades the field. (The biopsychosocial approach was not mentioned in the above discussion because it emerged out of medicine, and its strengths and weaknesses as a unifying approach are examined below.)

Is it Time to Leave Behind the Pre-paradigmatic Era of Psychology?

When he published *The Structure of Scientific Revolutions* in 1962, Kuhn argued that psychology and the other social sciences were pre-paradigmatic because a single perspective that was viewed as successfully explaining phenomena in these fields was lacking. Developments in the field since then continued to support that conclusion. Disagreements and conflicts between theoretical camps, researchers and practitioners, quantitative and qualitative researchers, and those with different views on empirically supported treatments have often been intense. In 1974, the National Council for Schools of Professional Psychology broke off from the APA due to fundamental differences regarding the appropriate training model for professional psychology education. In 1988, a large number of psychological scientists became disillusioned with the practice emphasis of the APA and broke off to form the American Psychological Society (the name was changed to the Association for Psychological Science in 2006). Controversies surrounding recovered memories of child sexual abuse became so intense in the 1990s that they became known as the "memory wars," one of the most conflictual periods ever in the history of psychology (Loftus & Davis, 2006). These various developments led many prominent psychologists to express concern that the field had become so fractured and the conflicts so intense that it might not be able to continue as a scholarly discipline (e.g., Buss, 1995; Gardner, 2005; Kendler, 2002; Rychlak, 2005; Staats, 2005; Sternberg, 2005; Vallacher & Nowak, 1994).

Fortunately, the intensity of these conflicts appears to be diminishing in recent years (Cummings, 2005; Goodheart & Carter, 2008; Magnavita, 2008; Norcross, 2005). Although psychology remains highly fragmented, there are signs of rapprochement in several areas. Within psychological practice, for example, there has been a trend away from relying on an individual unitary theoretical orientation and toward integrative approaches that incorporate multiple treatment methods (e.g., Norcross, 2005; Prochaska & Norcross, 2010). As larger numbers of psychologists employ integrative approaches, allegiance to the individual unitary approaches weakens and competition between their theoretical camps also diminishes.

There are other signs that the balkanization of the field is weakening. With regard to psychological science, the interconnected nature of research findings from across subfields and across the scientific disciplines generally is increasingly being recognized. Sternberg and Grigorenko (2001) argued that psychology must integrate findings from across the disciplines that investigate different aspects of human psychology. The importance of connection and consistency with the next levels of natural organization before a theory can be scientifically viable is increasingly being recognized (e.g., the necessity of evolutionary and neuroscience as well as sociocultural support for viable theories of psychological development, functioning, and behavior change; American Psychological Association [APA], 2003; Buss, 1991; Confer et al., 2010; Kaslow et al., 2007; McAdams & Pals, 2006; Sue & Sue, 2008).

The move toward evidence-based practice has become pervasive across health care and has become a priority within professional psychology as well. Though initially very controversial and divisive (e.g., the debate regarding the APA Division 12—Clinical Psychology empirically validated treatment effort was initially quite intense), there is now growing consensus that accountability and evidence-based practice will be enduring features of psychological practice (APA Presidential Task Force on Evidence-Based Practice, 2006). Norcross, Beutler, and Levant (2006) noted that the field is coalescing around a definition of evidence-based practice that integrates evidence from scientific research, clinical expertise, and patient values. This movement directs attention away from competing claims based only on theoretical arguments and toward the strength of the available evidence and the methodological improvements needed to increase confidence in research findings. For generations, it was acceptable for psychotherapists to appeal to theoretical support and personal experience alone when justifying one's approach to clinical practice. It is now the expectation that one needs to also point to empirical support for the effectiveness of interventions and the validity of the scientific rationale behind those interventions.

The competency-based education movement in the United States is also forcing professional psychology to re-examine its training philosophy and the goals and objectives used to prepare students for entry into the field. Though the subject of the competencies necessary for professional practice was very controversial in past decades (e.g., the fracture between those supporting the scientist-practitioner vs. the professional model and the breaking away of the National Association of Schools of Professional Psychology), a consensual approach to professional psychology education and training is now viewed as important to the continued development of the profession (e.g., Kaslow, 2004; Nelson, 2007). Education in the United States has moved decisively in the direction of developing competencies as opposed to completing coursework and other program requirements (Nichols & Nichols, 2001). This is forcing professional psychology, like other fields, to reconceptualize its training models and goals as it identifies the specific competencies needed to be proficient in the profession. Though consensus on what those competencies are has not yet been reached, extensive discussions have clarified the nature of the problem and the issues involved (Kaslow et al., 2004; Kaslow et al., 2007; Lichtenberg et al., 2007).

The technological and conceptual tools that are now available for conducting psychological science are also developing rapidly. In the past, it was not possible to directly investigate brain function or the role of genetics and other biological influences on psychological development and functioning, but research into these types of questions is now progressing steadily. Awareness of sociocultural influences on development and behavior is also steadily increasing (APA, 2003, 2007). In addition, methodological improvements in clinical research are also raising expectations for the validity of the evidence needed to support research hypotheses. Replicated findings from well-controlled studies (e.g., randomized clinical trials) and meta-analyses of findings from across studies are now expected before treatments are judged to be safe or effective (Chambless & Hollon, 1998; Wampold,

Lichtenberg, & Waehler, 2002). Problems with traditional approaches to psychological research using null hypothesis testing became widely acknowledged in the
1990s (Wilkinson & the Task Force on Statistical Inference, 1999), and these
approaches are gradually being replaced by mathematical modeling, which more
accurately reflects the nature of scientific inquiry (Rogers, 2010). These developments are rapidly shifting debates in the field toward much more sophisticated
analyses of psychological phenomena.

There also appears to be growing appreciation of the interconnectedness of living systems and levels of natural organization among scientists generally. One of
the most elegant descriptions of this perspective was offered by E. O. Wilson
(1998), the eminent biologist, who argued for *consilience*, or a "united system of
knowledge" (p. 298) across all the natural and social sciences and humanities.
Scientists are increasingly emphasizing these connections in their descriptions of
complex adaptive systems (e.g., Capra, 2002; Lovelock, 1979; Mitchell, 2009;
Wilson, 1998; Wolfram, 2002).

The impact of these developments is being felt across education, practice, and
research in professional psychology. In just a matter of years, the questions being
asked as well as the specificity and thoroughness of the answers being offered have
changed dramatically. Without a historical perspective, it may not be possible to
judge whether these developments represent a significant transition for the field.
Nonetheless, the confluence of these developments suggests the field may be entering a new era. Therefore, now is a good time to re-evaluate the adequacy of current
conceptualizations of education and practice in the field and to consider an intentional move to the next stages in the evolution of the profession.

This volume argues that it is now time to move past the problems associated
with the current assortment of theoretical orientations in professional psychology
and leave behind the pre-paradigmatic era of the field. The developments described
above have already reshaped the nature of psychological science and practice.
Not all the conceptual frameworks for structuring and organizing education and
practice within the field, however, have caught up. It certainly would be a tremendous relief to leave behind the perennial conflicts and controversies associated with
the competing traditional theoretical orientations and direct time and energies in a
more collaborative fashion on new and important questions and challenges. But to
leave behind the pre-paradigmatic era, a new comprehensive, unified framework is
needed that can replace the traditional assortment of theoretical orientations.
Before the adequacy of any such framework can be evaluated, however, it is important to first consider the basic issues that it must successfully address. Replacing
outmoded practices is certainly a worthwhile endeavor, but finding a satisfactory
replacement must be done responsibly and carefully.

Basic Requirements for a Paradigmatic Conceptual Framework for Psychology

The expansive nature of psychological practice provides a useful perspective for
evaluating the basic requirements that a unified paradigmatic conceptual framework

for professional psychology would need to meet. Psychologists work with an extraordinarily diverse range of individuals who experience a wide range of interacting psychological, sociocultural, and biological influences on development and functioning. In addition, a holistic perspective must be applied to develop an integrative understanding of individuals' development, functioning, and behavioral health treatment.

A unified conceptual framework for professional psychology would need to capture and reflect the full complexity of behavioral health care practice in order to survive clinical and scientific scrutiny. Specifically, it would need to address the tremendous complexity of human psychology so that it is applicable across the whole field, including all the different general and specialized areas of practice that address the full diversity of behavioral health concerns and biopsychosocial circumstances. It would also need to accommodate the research findings regarding the effectiveness of the full range of empirically supported interventions that have been identified. The strength of the scientific support for the framework would also need to be sufficient to provide the justification and rationale for clinical intervention. A theoretical framework that does not meet these basic requirements will likely fail as a unified paradigmatic framework for the field. Each of these issues is discussed below.

Complexity of Human Psychology

A requirement for a unified conceptual framework for professional psychology would be the ability to capture and represent the tremendous complexity of human psychology. As was noted earlier, the human brain appears to be the most complicated system known to exist in the universe. The range and capacity of human thought, emotion, and behavior is truly extraordinary and stretches even the capacity for comprehension. Nonetheless, psychologists would probably be unanimous in requiring that any unified conceptual framework for professional psychology must be able to capture this tremendous complexity.

There are signs of consensus regarding the basic characteristics of a theoretical framework that can represent this complexity. At minimum, such a framework needs to take a comprehensive approach that integrates biological and sociocultural factors into conceptualizations of psychological functioning. Virtually any of our textbooks for learning psychological assessment, along with our practice guidelines, accreditation and licensure standards, and the standards of practice identified by disciplinary bodies and malpractice courts, emphasize that psychological, biological, and sociocultural considerations all need to be incorporated into psychological assessments and evaluations. The same is true of treatment planning (e.g., see APA, 2002, *Ethics Code* 2.01(b); APA, 2003, multicultural guidelines; APA, 2006, *Whole Person Statement*; APA, 2007, *Guidelines for Psychological Practice with Girls and Women*; Eysenck, 1997; Joint Commission for the Accreditation of Healthcare Organizations, 2006, *Provision of Care Standards* PC 2.10−2.110; Kaslow et al., 2007; Melchert, 2007). Indeed, these guidelines and standards also suggest that failing to take a comprehensive approach such as this can result in incomplete case conceptualizations that can be ineffective and even deleterious.

Therefore, there does appear to be consensus that a unified conceptual framework for professional psychology that recognizes the highly complex nature of human psychology must take a comprehensive approach that integrates psychological, biological, and sociocultural influences into conceptualizations of human development and functioning.

Applicability Across All of Professional Psychology

A second requirement for a unified conceptual framework for professional psychology would be its applicability to the whole field of professional psychology, including all the general and specialized areas of practice. Contemporary standard textbooks presenting the traditional theories of personality, psychopathology, and psychotherapy normally note their varying applicability across different demographic and diagnostic populations. To the extent that a theoretical orientation cannot describe and account for behavior and functioning across practice areas, it is unlikely to be able to provide a common, unified conceptual framework for the field as a whole. A unified framework needs to be applicable across all levels of psychopathology and psychological functioning, all ages and demographic groups, and the full range of psychological, medical, and sociocultural issues with which psychologists work.

Many Effective Treatments

Over recent decades, it has become clear that a variety of interventions are effective for realizing therapeutic improvement and behavior change. Interventions that focus on cognition, affect, behavior, biology, or interpersonal and family processes can all be effective in terms of improving psychological symptoms, distress, and/or functioning. This finding was supported by the milestone meta-analysis published by Smith and Glass in 1977, as well as virtually any comprehensive review of treatment effectiveness conducted since then (see Chapter 10). This would also be the expected conclusion if human psychology is indeed highly complex and psychological outcomes are multifactorially determined in general. Consequently, an acceptable conceptual framework for professional psychology must be able to accommodate the diversity of interventions that have been empirically demonstrated to be effective in the treatment of psychological disorders and concerns.

Strength of Scientific Foundations of Professional Psychology

The above three requirements for a paradigmatic theoretical framework are relatively straightforward—there is probably strong consensus that a framework that does not meet those requirements could not serve as the unified conceptual framework for all of professional psychology. There is another requirement, however, that is more complicated. This concerns the strength of the scientific support for a unified framework that could inform clinical practice across all of professional psychology.

It would seem obvious that any unified theoretical orientation or conceptual framework for professional psychology must be consistent with and supported by the current body of scientific knowledge regarding human development, functioning, and behavior change. This requirement is not as obvious as it might seem, however, because comprehensive, detailed explanations are not available regarding many important psychological phenomena. While some areas of human psychology are well understood from a scientific perspective (e.g., the basic mechanisms involved in sensation and perception), other areas are not understood nearly as thoroughly (e.g., the development of personality characteristics, the causes of psychopathology, the nature and measurement of intelligence, and the mechanisms responsible for behavior change). In general, basic processes with fewer inputs and outputs are better understood, while more complex processes with larger numbers of inputs and outputs are less well understood. This has long been a major problem for clinical practice because more complex phenomena are usually the focus of treatment. As a result, clinicians have often had no alternative but to rely on one of the traditional theoretical orientations to guide their practice, even if those orientations were viewed as providing only partial explanations of personality, psychopathology, and/or behavior change.

The question of whether the scientific foundations of psychology are strong enough to provide a firm foundation for professional practice gets at our very core identity as a profession. The importance of psychological science for informing clinical practice has always been central to the core definitions of the field. Although psychologists vary in weighing the balance of "art" and "science" in clinical practice, the science component is normally viewed as essential. The scientific foundations of the field have also recently been emphasized more strongly in terms of the evidence-based practice movement (APA Presidential Task Force on Evidence-Based Practice, 2006). This has been an important movement throughout health care but represents a significant step for professional psychology to explicitly emphasize the central importance of research evidence in clinical practice.

Questions regarding the strength of the scientific foundations of psychology also need to be considered within the context of rapidly accumulating scientific advances. The pace of scientific progress in several areas has been truly remarkable (e.g., genomics, cognitive neuroscience). As a result of the development and refinement of new conceptual and technological tools (e.g., mathematical modeling procedures, computer hardware, genetic sequencers, high-resolution brain imaging machines), researchers are now able to directly examine aspects of human development and functioning that simply could not be observed in the past. Important advances are occurring across all the biopsychosocial domains. For example, instead of speculating about possible links between current human characteristics and our evolutionary past, it is now possible to verify and disconfirm testable hypotheses regarding these connections (Confer et al., 2010; Lumsden, 2005). New magnetoencephalography imaging machines with much improved spatial and temporal resolution are allowing researchers, neurosurgeons, and other clinicians to examine neural processing within the brain in far greater detail than was possible in the past (Cohen, 2004). Research has advanced regarding many sociocultural factors as well. Large bodies of empirical findings have accumulated at many levels, from the nature and consequences of the infant—mother attachment bond (e.g., Cassidy & Shaver, 2008; Mikulincer & Shaver, 2007) to the impact of ethnicity and culture (e.g., Sue & Sue, 2008; Suzuki, Meller, & Ponterotto, 2008).

These rapidly accumulating scientific findings are steadily strengthening the scientific understanding of many aspects of human psychology. But are they sufficient for informing a unified framework for professional psychology practice as a whole? Is enough now known about psychological development and functioning to justify a general move away from the current system allowing choices of theoretical orientations for conceptualizing patient cases and toward a single, unified science-based framework for use with all patient populations? Certainly there is still much that is lacking in terms of detailed and comprehensive scientific explanations of many psychological phenomena, and particularly with regard to the more complex processes that are important in clinical practice. As a result, this last requirement for a unified theoretical framework for the field is the most unsettled of the four issues raised in this section.

Conclusions

This chapter attempted to explain the complicated development of psychology as an academic discipline and field of professional practice. When viewed in the context of the development of the sciences generally, the emergence of the large number of competing theoretical orientations within the field is not nearly as odd as it might otherwise seem. In addition, there are many signs that the field has reached the point where it is ready to leave behind its confusing and conflictual pre-paradigmatic past.

In order to leave behind its pre-paradigmatic past, however, a new paradigm is needed that can successfully replace the assortment of theoretical orientations that professional psychology has relied on throughout its history. Before considering whether such a paradigm exists, some of the most basic requirements of a unifying paradigm for the field were discussed. There is likely agreement within the field regarding the first three of these requirements: that it be able to capture and represent the tremendous complexity of human psychology, be applicable across all the subfields and specialty areas within the field, and be able to accommodate all of the empirically supported interventions that have been shown to be effective in clinical practice. The least settled issue concerns the strength of current scientific findings for explaining human psychology and informing clinical practice. The position of the author on this last issue will be considered next.

There has been steady progress in the understanding of a wide range of phenomena throughout psychology. Several areas of inquiry have moved beyond the descriptive stage to experimental tests aimed at the verification of explanations and the falsification of hypotheses. For example, psychophysiology, the neurosciences, and genetics have employed experimental research methods for many years, and there are few questions regarding the validity of adequately verified and replicated findings for many topics within those fields. Additional areas are becoming steadily more experimental as well, producing increasingly reliable and thorough explanations for gradually larger numbers of phenomena. When one surveys the existing scientific literature in many subfields of psychology (e.g., behavioral genetics, comparative psychology, evolutionary psychology, sensation, perception, neuropsychology, development, emotion, cognition, personality), one finds that a large number of phenomena are understood at least in outline form and many processes are understood in significant detail. Despite all that remains to be reliably described and explained, the amount that is known has been steadily growing in recent years.

As knowledge of human psychology continues to grow, at some point scientific explanations will reach (if they have not already) a tipping point where the amount that is confidently known is sufficient to justify a general transition to a unified science-based conceptual framework for the practice of psychology. Knowing when this point has been reached is difficult to judge because the transition of a profession from primarily an "art" to a "science" involves a complicated and lengthy trial-and-error process. It is not a dichotomous, "all or nothing" decision point. That is, the science underlying an applied field does not need to provide detailed and complete explanations of phenomena in order to provide reliable scientific underpinnings for applied purposes.

Consider the relationship of physics to engineering. Many important questions, from particle physics to cosmology, have not yet been answered. There exists no "grand unified theory" for physics, a single unified theory that explains all the known forces and matter in nature, the discovery of which has been seen as the ultimate goal for physics (Greene, 1999). Indeed, there is no consensus that a grand unified theory even exists (Mitchell, 2009). Nonetheless, the available knowledge in physics is sufficiently detailed and comprehensive to provide thoroughly reliable scientific foundations for a vast range of applied purposes in engineering. The public has few qualms, for example, about using skyscrapers, jet planes, MRI machines, and many other remarkably effective applications of the science of physics.

The same is true of the relationship of biology to medicine. Despite the remarkable progress of the biological sciences over the past century and in recent decades in particular, explaining health and disease often remains highly imprecise. There are many diseases for which both the cause and the cure are unknown (i.e., idiopathic diseases such as Alzheimer's, Parkinson's, multiple sclerosis, rheumatic arthritis, epilepsy). For many diseases, it is not possible to predict who will develop pathology, who will respond to treatment, and who will remain in remission. Nonetheless, scientific explanations of biological processes are sufficiently strong that nearly everyone is willing to rely on science-based medical and public health measures for dealing with physical health problems.

The field of medicine addressed the "tipping point" question when it sent Abraham Flexner to review the scientific foundations and quality of medical education across the United States in 1908. His conclusion (Flexner, 1910), that a large amount of medical education was not sufficiently science-based, led to a major reconceptualization of medical education in the United States and even to the closing of numerous medical schools (Hiatt & Stockton, 2003). And while many questions about human biological functioning still remained unanswered at the time of Flexner's report (e.g., penicillin was not discovered until 1928 and was not mass produced until the end of World War II), a unified science-based framework for practicing medicine was not viewed as premature. Professional psychology may be in a somewhat similar situation today. While many aspects of psychological functioning remain poorly understood at this point, some are at least partially understood and others are understood reasonably well. Therefore, applying a unified science-based framework to the practice of psychology may not be premature.

This volume argues that the tipping point has now been reached and professional psychology is ready to embrace a unified science-based approach to understanding human development and functioning. This conclusion does not imply that a "grand unified theory" for psychology has been reached. Given the status of the grand unified theory in physics, a field that is more than 2,000 years old, requiring similarly comprehensive scientific explanations in the young science of psychology would be wildly premature (see Crutchfield, Farmer, Packard, & Shaw, 1986; Gordon, 2007; Mitchell, 2009). This conclusion that the tipping point in psychology has now been reached, however, does imply significant confidence in the strength of current scientific findings for explaining human psychology. The rationale supporting this conclusion is explored further in the next chapter.

5 The Biopsychosocial Approach: General Systems, Nonlinear Dynamical Systems, and Complexity Theory

Since psychology became established as a scientific discipline in the late 19th century, theorists have argued that a variety of psychological, sociocultural, or biological factors are the primary determinants of development and behavior. Freud argued that early child-rearing practices mold the manner in which individuals manage the biologically based sexual instincts that all humans must deal with. He believed that the resulting patterns for managing these instincts were largely unconscious and highly resistant to change. Watson, on the other hand, took a very different view. Though he believed that genetic influences were important, he viewed behavioral conditioning as very powerful, as reflected in his famous statement, "Give me a dozen healthy infants, well-formed, and my own specific world to bring them up in and I'll guarantee to take any one at random and train him to become any type of specialist I might select—a doctor, lawyer, artist, merchant-chief and, yes, even into a beggar-man and thief, regardless of his talents, penchants, tendencies, abilities, vocations and race of his ancestors" (1925, p. 82). Others argued that cognitive beliefs, family system processes, biological factors, or a variety of other factors are the primary determinants of personality and psychopathology.

From a contemporary perspective, the role of these individual factors in human development and functioning was generally overstated. Though each offers an interesting perspective or insight into psychological phenomena, none provides a thorough explanation. If these theories could somehow be meshed together, however, they would provide a better approximation of the complexity of human psychology.

This is essentially the problem that George Engel addressed when he proposed the biopsychosocial model in 1977. Engel, a physician, had concluded that the biomedical approach, the prevailing conceptual framework for practicing medicine and psychiatry at the time, was inadequate for capturing the complexity of medical illness. As he stated, "I contend that all medicine is in crisis and, further, that

Foundations of Professional Psychology. DOI: 10.1016/B978-0-12-385079-9.00005-9

medicine's crisis derives from the same basic fault as psychiatry's, namely, adherence to a model of disease no longer adequate for the scientific tasks and social responsibilities of either medicine or psychiatry" (p. 129). Engel viewed an exclusively biomedical approach to understanding disease and its treatment as reductionistic, ignoring large amounts of scientific evidence regarding the critical role of psychosocial influences on physical functioning. He concluded that "To provide a basis for understanding the determinants of disease and arriving at rational treatments and patterns of health care, a medical model must also take into account the patient, the social context in which he lives, and the ... physician role and the health care system. This requires a biopsychosocial model" (p. 132).

Engel (1977) argued that the etiology and/or treatment of many medical and psychiatric problems included large psychosocial components and that the "psychobiological unity of man" (p. 133) needed to be integrated into assessment and treatment planning in order for medicine and psychiatry to become more effective at treating illness and promoting health. Engel argued that "the physician's basic professional knowledge and skill must span the social, psychological, and biological, for his decisions and actions on the patient's behalf involve all three" (p. 133). The basic definition of the biopsychosocial approach used in this volume is the same as that advocated by Engel. It views humans as inherently biopsychosocial organisms in which the biological, psychological, and sociocultural dimensions are inextricably intertwined (Melchert, 2007). Consequently, behavioral health care, prevention, and health care generally all need to be based on this basic premise.

This chapter reviews the scientific basis underlying the biopsychosocial and other approaches to understanding complex systems. It begins by describing the conceptual model underlying Engel's biopsychosocial approach, namely von Bertalanffy's (1968) general systems theory—the approach to complexity theory most familiar to professional psychologists because of its influence in family therapy. It then introduces other approaches to nonlinear dynamical systems and complexity theory, of which general systems theory is a part. Though complexity theory approaches are not widely known within psychology, they have been used for investigating the nature of complex systems across the sciences for several decades. The human being is the archetypal example of a complex system, and so these approaches need to be incorporated into the scientific foundations of professional psychology. The purpose of this introduction to complexity theory is to convey how well established these approaches are in the sciences and their usefulness for understanding the tremendous complexity of human psychology that has been difficult to capture using the traditional psychological theoretical orientations.

Introduction to General Systems, Nonlinear Dynamical Systems, and Complexity Theory

George Engel based the biopsychosocial model specifically on von Bertalanffy's general systems theory. Starting in the 1950s, the biologist Ludwig von Bertalanffy was interested in "the formulation and deduction of those principles which are valid for 'systems' in general" (1950, p. 139). Von Bertalanffy defined *system* in a very general sense as a collection of interacting elements that together produce some form of system-wide behavior. Most of his interest was in identifying properties of living systems, ideas that were then more fully explained in his very influential 1968 book, *General System Theory*. He argued that models of *closed systems* used to explain nonliving physical phenomena were not applicable for living systems. His emphasis on *holism* and his descriptions of *open systems* that are dependent on information, feedback, and communication were highly influential across the physical and social sciences (Davidson, 1983).

Simon (1962) extended the understanding of systems by focusing on the nature of subsystems within the larger system. He argued that one of the most important attributes of a system is its *degree of hierarchy*. Living systems in particular are structured hierarchically—a body is composed of organs, which are composed of cells, which are composed of cellular subsystems, which are composed of molecules, and so on. The complexity of a system can then be measured in terms of its hierarchy. The other important attribute of systems are their *near-decomposability*, by which Simon meant that a subsystem has much stronger interactions within its subsystem than it has with other subsystems. As a result, subsystems function with a significant degree of independence (e.g., an organ has much stronger interactions within its internal components than it does with other organs, as do members of a family). Simon argued that complex systems are able to evolve in nature as a result of these two characteristics.

Along with cybernetics, general systems theory was highly influential in stimulating the development of many fields including systems biology, systems ecology, artificial intelligence, neural networks, control theory, and other areas of science and engineering (Mitchell, 2009). Within psychology, neural networks have become critical to the modern understanding of the mind and brain, and systems theory provided the main theoretical underpinnings for the development of the family therapy field. As will be seen below, there are also many other applications of general systems and other complexity theory approaches to understanding psychological processes.

General systems theory has been very influential across the sciences, but it is just one of many approaches to understanding complex natural phenomena. These various approaches are subsumed under the general category of *nonlinear dynamical systems theory*, which is often referred to as *complexity theory* when the focus is on systems with more complexity (i.e., more variables; Gros, 2008). Though the mathematical expression of nonlinear dynamical systems and complexity theory concepts is typically highly complicated, the basic concepts involved can be intuitively quite appealing because they often seem to capture the complexity of lived experience. They are typically not taught in undergraduate or graduate psychology programs, however, and so many psychologists are not familiar with them. Therefore, before introducing basic aspects of complexity theory, the next section notes examples of higher order complexity that are familiar to psychologists and reflect the importance of nonlinear and complexity theory concepts for understanding psychological phenomena.

Familiar Examples of Complex Systems for Psychologists

Psychologists are obviously familiar with highly complex systems. A child's development, the nature of psychopathology, the functioning of families, groups, and organizations, and essentially all of the phenomena that psychologists work with involve highly complex systemic phenomena. Most of the traditional research methodologies that psychologists learn were not designed to reflect this complexity, however. Noting some of the limitations of these traditional methodologies points to the need for adding nonlinear and complexity theory approaches to the traditional ones.

The Nature of Change

Life inherently involves change. Sometimes things change in a predictable straightforward manner, but at other times they change only very gradually, and at other times they change dramatically. In the simplest form of linear change, outcomes are directly proportional to inputs in a straightforward manner that is characterized by the familiar bivariate regression equation, $Y = a + bX$, where a change in one parameter (or some linear combination of parameters in multiple regression) results in a corresponding change elsewhere in the system (i.e., every time X increases by one, Y increases by b). The ability to make predictions in this manner is extremely powerful and accounts for a great deal of the success of science historically. But much of nature does not operate in this simple, linear manner.

When it comes to living systems, change is typically very complex. Sometimes no change occurs, as in states characterized by equilibrium or homeostasis (Cannon, 1932). At the other extreme is complete randomness, where changes are completely chaotic or unpredictable (at least according to any known explanations). In between are myriad other types of change. Even several multiple regression analyses combined in a structural equation model are not able to capture all these complex change processes quantitatively. For example, developmental psychologists have embraced the concepts of equifinality (where multiple developmental paths can lead to the same outcome) and multifinality (where the same developmental pathways can lead to multiple different outcomes; Ollendick, 2008). These common types of change processes cannot be modeled using the general linear model that has been the mainstay for statistical analysis in psychology (Rogers, 2010). Nonlinear mathematical modeling is required instead.

Statistical Interaction

Another familiar example of the complexity of change involves statistical interaction. In its simplest three-variable case, the relationship between two variables is modified by the effect or value of the third variable. The discovery of interactions in psychological research is often highly welcomed because of their explanatory power regarding the relationship between two or more factors.

In the psychological and the living world in general, however, interaction is typically highly complex and involves a very large number of variables. Very few outcomes are caused by single variables (e.g., eye color is one of the relatively few human characteristics that are genetically determined in a simple, straightforward fashion through the interaction of just two genes). Instead, multiple factors from across the biopsychosocial domains typically interact in causing phenomena, and these causes generally do not interact in a simple summative fashion. Factors sometimes cancel each other out, reinforce each other in nonlinear ways, or interact in higher-order ways. Large inputs frequently produce small effects, while a small input at the right time can produce a very large effect. As a result, the whole truly is greater than the sum of its parts—something the general linear model of statistics is not designed to capture. Structural equation modeling has become very influential in behavioral science research because it can capture the effect of higher-order interactions by creating new variables that represent the interaction of measured variables. While this type of modeling advance represents the complexity of phenomena more comprehensively, its ability to do so is nonetheless limited as a result of measurement error, the difficulty of identifying all of the influential influences on phenomena (i.e., inadequate specification of the model), and the inability to capture higher-order nonlinear change processes. There are many other kinds of modeling procedures available, however, for capturing different types of change processes (Gros, 2008; Guastello & Liebovitch, 2009; Rogers, 2010).

Psychometrics

Another illustration of complexity familiar to psychologists involves the distinction between the classical test theory of psychometrics and new item response theory and generalizability theory approaches. Classical test theory involves differentiating the variance in scores attributable to one's true score from all the rest of the variance, which is then attributed to random error ($X = T + E$, or *Observed score = True score + Error*). This is similar to the purpose of much psychological research in general, where the purpose is to identify the variability that is due to the model while the remainder is then attributed to error. Complexity theory, however, is conceptually closer to the newer psychometric approaches. Generalizability theory, for example, examines the "facets" or sources of variation in test scores that are attributed to random error in the classical theory but may be associated with persons, items, settings, time, and other factors. The purpose of generalizability theory is to quantify the amount of error caused by each facet and their interactions. Though item response and generalizability theory rely on linear conceptualizations, they are able to examine more of the complexity behind the scores obtained by individuals taking psychological and educational tests.

Newtonian Mechanics Versus Thermodynamics

Another clear example of the difference between simpler linear phenomena and more complex systemic behavior involves the difference between Newtonian mechanics and thermodynamics. Newtonian mechanics focuses on the behavior of particular objects, including their position, velocity, and trajectory: for example, the acceleration of the legendary apple that fell from a tree and hit Newton on the head. Thermodynamics, on the other hand, is concerned with the behavior of whole systems and the impact of the environment on the behavior of the system. For example, understanding the nature of heat involves examining systemic characteristics such as pressure, volume, and temperature—the behavior of the thermodynamic system cannot be understood without examining properties of the whole system. It is not possible to predict the behavior of a thermodynamic system by examining the linear behavior of each molecule. Though in theory it may be possible, it involves far too many computations to be practical at least in the foreseeable future—instead, its systematic properties must be examined (Capra, 1996).

Definition of Nonlinear Dynamical Systems and Complexity Theory

To understand the value of particular nonlinear dynamical system or complexity theory approaches for understanding the nature of complex systems, some basic definitional and conceptual background needs to be reviewed. First of all, "there is no one agreed-on quantitative definition of complexity theory" (Mitchell, 2009, p. 13). There also is no one single *complexity theory* or *science of complexity* (Mitchell, 2009). Though at first this might seem to be a significant problem, this is actually not unusual in science. There is not one single definition for many important areas of study, such as quantum mechanics, genetics, or, in psychology, intelligence or self-concept.

Conceptually, however, there is clear agreement regarding the nature of nonlinear dynamical systems and complexity theory. Mitchell (2009, p. 15) has offered the following definition: "Dynamical systems theory (or *dynamics*) concerns the description and prediction of systems that exhibit complex *changing* behavior at the macroscopic level, emerging from the collective actions of many interacting components. The word *dynamic* means changing, and dynamical systems are systems that change over time in some way." She further notes that complex systems change and adapt their collective behavior based on information from both their internal and external environments. This occurs at many levels from the single cell, an organ (e.g., the brain), human social interaction, the economy, the weather, the biosphere, and all the way up to the cosmos. The nature of change at all these levels is typically highly complex and nonlinear. Indeed, as Stewart (1989) put it, nature "is relentlessly nonlinear."

Dynamical systems theory is also used to refer to adaptive, changing (i.e., dynamical) systems that contain a limited number of variables, while complex systems theory focuses on dynamical systems containing very large numbers of variables (Gros, 2008). Both use coupled differential equations to model the nature of phenomena, but the former typically contain fewer variables.

The complexity of life has always been recognized, of course. Locke (1690/1975, p. 2) wrote that "Ideas thus made up of several simple ones put together, I call Complex; such as are Beauty, Gratitude, a Man, an Army, the Universe." But this complexity was simply far too complex to study in a quantitative scientific manner using the relatively rudimentary methodological tools that were available at the time. Over the past century, however, several technological and conceptual tools have been developed that allow a quantitative investigation of complex phenomena. And, of course, the most challenging of phenomena for investigation is the human mind and brain. As Gros (2008, p. 181) noted, "The brain is without doubt the most complex adaptive system known to humanity, arguably also a complex system about which we know very little." Understanding complexity in general is also among the leading goals for science. Stephen Wolfram, one of the world's leading current scientists, observed that "what's perhaps the most long-standing mystery in all of science: where, in the end, the complexity of the natural world comes from" (quoted in Malone, 2000).

Nonlinear Dynamical Systems and Complexity Theory as Metatheory

There is one more critical point that needs to be appreciated before specific dynamical systems and complexity theory approaches are introduced. The general terms discussed thus far—the biopsychosocial approach, general systems theory, nonlinear dynamical systems theory, and complexity theory—are normally used to refer to *metatheories* rather than scientific theories *per se* because they are not scientific theories that provide explanations for specific psychological, biological, or social phenomena. Instead, they provide the overarching conceptual perspective used to gain a comprehensive understanding of phenomena (Guastello & Liebovitch, 2009). Theories in the scientific sense refer to a principle or body of principles that explain a class of phenomena. Using empirical observations, rules, and scientific laws, they describe current observations and predict future observations of phenomena (Hawking, 1996; Popper, 1963). A metatheory, on the other hand, is "a theory the subject matter of which is another theory" (*Encyclopaedia Britannica*, 2010). In other words, it is a theory about other theories. Metatheories do not explain the behavior of specific phenomena (e.g., the development of specific disorders or the nature of particular cognitive processes). Instead, they are necessary for explaining the nature of complex phenomena that result from the interaction of many specific processes. Metatheories provide the general conceptual framework indicating the essential characteristics that need to be integrated to understand theory and research in particular areas of scientific or philosophical inquiry.

This volume refers to complexity theory and the biopsychosocial approach in this metatheoretical sense. It is this overarching perspective that is able to provide the unifying framework for professional psychology. Specific scientific theories accompanied by their corresponding experimental literatures are needed to explain specific psychological processes. An overarching metatheoretical perspective is also needed, however, for understanding complex phenomena comprising many interacting processes. This is true not only with human psychology but with biology as well as physics. Despite the large amount of scientific knowledge regarding biological and physical phenomena, there are no unified theories of biology or physics at the present time—single theories that explain all the natural phenomena in the physical or the living, biological worlds. It is impossible to say when unified theories of physics, biology, or psychology will be developed and verified, or even whether such unified theories exist (Mitchell, 2009). In the meantime, however, we need to rely on metatheoretical frameworks to organize and integrate available scientific knowledge within and across these sciences. All of the sciences operate similarly in this manner.

Historical Origins of Complexity Theory: Chaos Theory

Complexity theory in the sciences is roughly the same age as scientific psychology. Mitchell (2009, p. 21) notes that Henri Poincare was "the founder of and probably the most influential contributor to the modern field of dynamical systems theory." In 1887, he attempted to solve a problem concerning the stability of the planets in their orbits around the sun. (Just a few years earlier, in 1879, Wilhelm Wundt had established the first psychology research laboratory in Leipzig, Germany.) There was interest at the time in the question of whether the rotation of the solar system might be unstable and eventually spin out of control. Poincare discovered that miniscule differences in the initial conditions of objects (here, in the initial positions, masses, or velocities of the planets) can produce very large differences in the later motion of those objects and the system in general. He later wrote that "it may happen that small differences in the initial conditions produce very great ones in the final phenomenon. A small error in the former will produce an enormous error in the latter. Prediction becomes impossible" (Poincare, 1914).

The phenomenon that Poincare discovered required highly complex mathematical processing to solve, however, and could not be proven until the invention of the computer many decades later. But it has since become an accepted scientific fact that has become known as *chaos theory*. The first major empirical demonstration of this effect was accidentally discovered by Edward Lorenz, a meteorologist. Lorenz (1963) discovered the effect when he reran a weather modeling program with slightly different input data. To save time and paper, he reran a particular program in the middle of the sequence instead of the beginning, using a starting value of 0.506 instead of 0.506127 (i.e., he rounded off the original value, thinking such a miniscule difference would have no practical effect). When he returned to the lab, he was amazed to find a completely different result (Baofu, 2007). The difference in the accuracy of the starting value was so small that it was comparable to the force of a butterfly beating its wing, and the effect consequently became known as the "butterfly effect." The eventual effects produced by such exceedingly small variations in input were so great that they became known as *chaotic*. Probably the best-known example of this effect are meteorological predictions which generally become quite unreliable (chaotic) after a week or 10 days.

Describing these phenomena as "chaotic" could be considered an exaggeration, however, in that outcomes are typically not completely random and indeterministic.

The weather is predictable in the short run and climate is predicable in the long run. In a system of absolute chaos, on the other hand, anything could happen— small changes in one part of the system could produce indeterminate results. If chaotic effects such as this were found throughout the natural world, of course determining causality and predicting outcomes would be impossible. In principle, it is possible to measure inputs with such accuracy that outcomes, even in cases of highly complex interactions between inputs and outcomes, could be predicted accurately (albeit in nonlinear terms). In practice, however, attaining this level of measurement and building such complex models of systems is extremely difficult and may never be possible for the most complex of systems (Gros, 2008).

As everyday experience clearly suggests, small changes in one part of the system typically results in *some range of* possible outcomes. A butterfly's wing beat in China might result, under the right conditions, in violent storms on the west coast of the United States. It will not result in the United States becoming a tropical rainforest or an Arctic permafrost, however, unless the weather system is at a bifurcation point (see below). That is, the outcomes of small perturbations in input parameters result in determined, organized chaos such that certain things are likely to result, though it is not possible to predict which one (Byrne, 1998).

Mitchell (2009) illustrates the basic nature of chaotic systems by contrasting linearity and nonlinearity. She notes that when baking, for example, adding two cups of flour and one cup of sugar results in three cups of ingredients in a linear fashion—"the whole is equal to the sum of the parts" (p. 23). When a cup of vinegar is added to two cups of baking soda, however, the result is a much larger amount of vinegar, baking soda, and carbon dioxide fizz that results from the nonlinear (and violent) interaction of the vinegar and baking soda. The nonlinear result involves a far more complex interaction (and far more complex mathematics to model). The difference in complexity between many linear and nonlinear processes is often very large, prompting Mitchell (2009, p. 23) to note that "Linearity is a reductionist's dream, and nonlinearity can sometimes be a reductionist's nightmare."

"Logistic Map" and Attractors

Most of nature involves nonlinear processes (Capra, 1996; Stewart, 1989), and capturing these processes normally involves highly complex mathematics to model (for introductions, see Gros, 2008; Guastello & Liebovitch, 2009). Mitchell (2009) notes that a very simple equation, however, is able to capture the very essence of chaos—the sensitive dependence of change on initial conditions. This simple equation, $x_{t+1} = Rx_t (1 - x_t)$, has become known as the *logistic map* and is "perhaps the most famous equation in the science of dynamical systems and chaos" (Mitchell, 2009, p. 27; the equation was first published by Verhulst in 1845 but did not become famous until the 1970s). When the value of R in this equation is changed, the result varies in dramatic ways. When x at the present time (or x_0) is 0.2, for example, and R is 2.0, then the result quickly reaches a fixed point and stays there. At time 5 (or x_5), the result is 0.5 and it does not change further—it remains at 0.5 into the future. Even when x_0 is increased but R remains at any value less than 3.0, the result still reaches 0.5 (though it takes a longer and more convoluted path to get there). This is known as a *fixed point attractor* because the result is "attracted to" a fixed point and then stays there.

Attractors operate like a magnet, pulling in objects unless strong enough forces overpower the attractor. A pendulum also illustrates the operation of a fixed-point attractor (Byrne, 1998). A swinging pendulum will eventually stop due to the force of gravity at the attractor point, which is the bottommost point of the pendulum (attractor) basin, given the very limited number of control parameters that act on the pendulum. Various forms of the logistic map and other descriptions of attractors have been used to describe the nature of change in many physical and biological phenomena. Within psychology, fixed point attractors have recently become influential in the conceptualization of personality, social interaction, neural processing, and a variety of other psychological phenomena (see below).

Bifurcation

When R in the logistic map equation is increased to 3.0, however, the result changes dramatically, settling into an oscillating pattern between two distinct values. This point at which the oscillation occurs is called a *bifurcation*. When the result bifurcates into two repeating values, the change process is known as a *periodic attractor* because the ultimate values are "attracted to" two periodically varying values between which they then oscillate (known as a period-two attractor). When R is increased to ~ 3.45, the resulting values bifurcate again and oscillate between four distinct values (a period-four attractor). When R is increased to ~ 3.54, another bifurcation occurs, resulting in an oscillation between eight distinct values (a period-eight attractor). Feigenbaum (1980) discovered that when R reached ~ 3.5699, the number of bifurcations that occurred was infinite, which denotes the onset of chaos. *Feigenbaum's constant* describes an essential feature of this period-doubling cascade toward chaos, an effect that has been verified in laboratory examinations of electronic circuits, lasers, fluid flow, and chemical reactions. More complex phenomena have been too difficult to examine directly, but computer models of the weather, electrical power systems, the heart, and many other systems have also shown this period-doubling route to chaos. Feigenbaum's constant is consequently known as a universal feature of the many complex systems that exhibit chaotic features (Mitchell, 2009).

Bifurcations provide another illustration of the difference between linear and nonlinear processes. Psychologists are familiar with the law of large numbers, the consequence of which is that when sample sizes become large, the effect of fluctuations in sampling becomes small. This principle is critical when conducting research based on the general linear model of statistics. In complex systems, however, small fluctuations are critical because they can have dramatic effects when they near bifurcation points (Prigogine & Stengers, 1984). When these transformation points are reached, systems take different trajectories based on very small differences. Systems in equilibrium can accommodate fluctuations in their control parameters, but if perturbations in these parameters reach a certain point, a system will destabilize at the bifurcation point and will fluctuate between two or more new points, eventually taking on a new path in its development (Harvey & Reed, 1994).

A familiar example of this effect involves the world's climate. Small fluctuations in weather do not affect the overall pattern of the world's climate when the system is in equilibrium. Small perturbations in weather at bifurcation points in the world's geologic history, however, have led to dramatic changes in climate such as the Ice Age and the later return to warm temperatures. (The recent rise in global temperatures of course gives urgency to gaining a better understanding of bifurcation points in climate systems.)

In psychology, children's development provides a good illustration of the importance of bifurcations. In stage models of development such as Piaget's, a particular stage can be thought of as a fixed point attractor—a relatively stable state that characterizes many complex systems (van Geert, 2009). These states can also be thought of as being in equilibrium or homeostasis (Vallacher & Nowak, 2007). During their development, however, children also make qualitatively different, discontinuous jumps in ability (e.g., in terms of reaching and grasping, language, reasoning). Continuous linear models of development do not capture the nature of these jumps. The concept of bifurcation, on the other hand, describes this phenomenon well—when children reach bifurcation points in their cognitive development, they can jump to a qualitatively different level of ability or functioning very quickly, and then remain in a relatively stable new state for a significant amount of time. A child's development is considerably more complex than what is suggested by this simple example of bifurcation (i.e., a child's environment has a very important impact on his or her development, as do biological factors), but it nonetheless illustrates how nonlinear concepts such as bifurcation are necessary for conceptualizing developmental processes (Lunkenheimer & Dishion, 2009; van Geert, 2009). Indeed, understanding evolution in general requires these types of concepts. For example, the evolutionary record consists of *punctuated equilibria* where long homeostatic periods are interspersed with dynamic periods characterized by the extinction of certain species and the emergence of others, events that often also correspond with shifts in the global ecosystem (Bak, 1996). Thus, both children's development and the evolution of life in general represent prime examples of complex, nonlinear dynamical systems that cannot be explained through linear models of development and change.

Other Important Nonlinear Dynamical Systems Concepts

There are several important methodological and theoretical approaches to nonlinear dynamical systems and complexity theory besides chaos theory that are important for understanding human psychology. Three additional concepts important to a basic introduction to complexity theory are fractals, self-organization, and emergence. (For more extensive introductions to the conceptual basis of complexity theory, see Capra, 1996, Mitchell, 2009; for an introduction to the mathematics of complexity, see Gros, 2008, Guastello & Liebovitch, 2009; for a review of applications in psychology, see Byrne, 1998, Guastello, Koopmans, & Pincus, 2009, Vallacher & Nowak, 2007.)

Fractals

A defining characteristic of fractals is that they are self-similar, which means that their small parts are like their larger parts. In nature, self-similarity is not perfect: a tree branch looks similar to (but not exactly like) the whole tree, the end of a branch looks similar to the whole branch, and a leaf shares similarities to the end of the branch, though it does not look exactly like it (see Figures 5.1 and 5.2). Clouds, snowflakes, roads, cities, and the coastlines of islands and continents all share this characteristic.

Fractals (or fractional dimensions) are geometric structures where the number of dimensions is not an integer (e.g., the number of dimensions can be anywhere

Figure 5.1 Fractal self-similarity in ferns: The structure and overall shape of the main branch in the Giant Rabbit's Foot Fern on the left is replicated in the structure of each small branch that emanates from the main branch, and each leaf again replicates the next larger structure and shape; the same pattern is also evident in the Anthurium fern on the right.

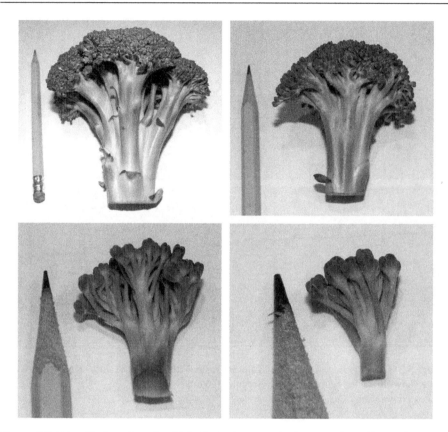

Figure 5.2 Brocolli show fractal self-similarity between the larger and smaller portions of the plant, from the whole stalk down to the smallest bud (each subsequently smaller portion was cut from the larger portion shown in the preceding photo).
(Photos courtesy of Jackson Melchert.)

between 1 and 2). A classic example of a fractal is the coastline. Consider the coastline of Great Britain, which is jagged when viewed from an orbiting spaceship, with its large peninsulas and bays. It looks similar when viewed closer, as from an airplane, though on a smaller scale. When viewed on foot, it is similarly rugged and jagged, though now on a much smaller scale. When one bends down to examine bugs crawling on rocks at the water's edge, it still shows similar kinds of jaggedness. The similarity of the shape of a coastline at different scales (its *self-similarity*) is a prime example of fractal geometry. Measuring the length of Great Britain's coastline greatly increases when it is examined at finer scales, and there is no one best measure of the length of the coastline—the length simply varies depending on the resolution with which it is measured. If the coastline were measured with a largely one-dimensional line, its length would always be the same no matter how it was measured. When it is measured on a much finer scale, its length is far greater (see Figure 5.3). The coastline is not so jagged, however, that measuring it at high resolution fills up a two-dimensional space. Therefore, it can be considered to have

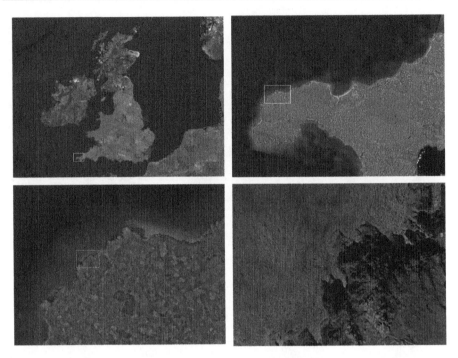

Figure 5.3 The coastline of Britain at different scales—though not exactly self-similar, each image shows the same types of jagged irregular features.
(Photos copyright by Google Earth.)

a dimension somewhere between one and two. Its high jaggedness results in it being closer to two, as compared with the coast of Africa, which is smoother and closer to one. Fractals can also have a time dimension, which is critical for understanding change in living systems and human behavior (Guastello & Liebovitch, 2009). The ability to describe self-similarity across time represents a major advance in the scientific understanding of natural phenomena and is particularly important with regard to understanding living adaptive systems.

Fractals have recently been found to underlie a number of important biological processes. If verified under a sufficient number of conditions, this development may represent a very significant paradigmatic change in the evolution of biology as a science. West, Brown, and Enquist (1999) used fractals to explain the Kleiber biological scaling law, the observation that large animals are able to support a higher metabolism than one would expect given their relatively limited surface area. (An animal's surface is needed to dissipate the heat caused by metabolism, but the surface area of increasingly larger animals increases much more slowly than does their mass. If not for this scaling law, large animals would die and actually start on fire due to the excessive heat caused by their metabolism.) This law has been observed not only in mammals and birds, but also fish and plants—indeed, all sizes of living organisms from bacteria to whales (Mitchell, 2009). The lifespan of animals also follows a similar fractal pattern: the larger an animal is, the longer its lifespan (though humans are one of the exceptions to the rule). The West et al. (1999) metabolic scaling theory has provided explanations for many other biological phenomena, including heart rate, gestation time, plant growth rates, and the rate of DNA changes over evolutionary time. Being a new theory, it will take years before it can be sufficiently tested under many different conditions. The initial results are impressive, however, and there is excitement among some scientists that it may be able to unify all of biology under one theoretical framework (Mitchell, 2009).

Complexity in psychology has also been observed to be fractal, and many examples of self-similarity in cognition, emotion, and behavior across time have now been found. Social interactions such as those that take place within family therapy, group therapy, and other interpersonal dynamics, as well as a variety of neural processing and cognitive processes such as emotion, reasoning, speaking, memory, and intentional behavior in general, have been shown to be fractal in nature (Guastello, Koopmans, & Pincus, 2009; Hollis, Kloos, & Van Orden, 2009; Minelli, 2009; Pincus, 2009). Indeed, fractals are good indicators of the complexity of systems in general (Guastello & Liebovitch, 2009).

Self-Organization

Throughout history, philosophers and scientists have viewed living systems as self-organizing networks where the subsystems and components are all interconnected and interdependent. Examples of this include the economy (cf. Smith's (1776) *invisible hand of the market* that causes the marketplace to self-regulate), the family, the individual organism, and the individual cell (a dramatic example is the ability of fertilized eggs and stem cells to grow into any tissue in the body). It has been possible to quantify the nature of self-organizing systems only recently, however, as the result of the development of complex mathematical modeling tools and the very large computational capacity of modern computers (Capra, 1996; Wood et al., 2006).

Self-organization allows systems that function in a diffuse, disorganized manner to operate more efficiently (Prigogine & Stengers, 1984). All forms of self-organization rely on information flows (e.g., feedback loops) of some type, which is consistent with John von Neumann's view that all of life is an expression of energy flow (von Neumann was one of the most influential scientists of the past century; Levy, 1992). Mitchell (2009) noted that many scientists now believe that information has become as important as mass and energy in understanding the fundamental components of reality, and that living systems in particular are essentially highly complex information-processing networks. Human psychology is a prime example of this perspective.

Emergence

Emergence refers to the way that complex patterns rise out from the multiplicity of simple interactions in complex systems. It is exemplified by the phrase "The whole is greater than the sum of its parts." The complex symmetrical patterns of snowflakes or the ripples in sand dunes or water caused by the wind are examples of emergence in physical systems. Examples of emergence in living systems can be highly complicated as the result of interactions between genetic inheritance, the environment, and time. Examples include the collective intelligence of ants (Sulis, 2009) or the emergence of leadership in human groups and organizations (Guastello, 2009). Schwartz and Begley (2002, p. 350) note that "An *emergent phenomenon* is one whose characteristics or behaviors cannot be explained in terms of the sum of its parts; if mind is emergent, then it cannot be explained by the brain."

How Effective Are Nonlinear Dynamical Systems and Complexity Theory in Psychology?

The physical sciences have employed nonlinear dynamical systems and complexity theory approaches for decades, and many of the recent advances in understanding complex natural phenomena could not have been achieved without employing these approaches. Advances in computational capacity and mathematical modeling over the past half-century have allowed for the verification of these approaches at very high levels of precision. Applying these concepts in the behavioral sciences has occurred more recently, and consequently there is far less empirical evidence regarding their accuracy and validity for explaining psychological phenomena. Complexity concepts have been very influential in psychology at the metatheoretical level but have not undergone the extent of quantitative testing as they have in the physical sciences.

Nonetheless, quantitative tests of complexity theory approaches in psychology are showing them to be highly useful and generally superior to linear approaches. One of the best known is Gottman's research on marital relationships. Gottman, Swanson, and Swanson (2002) applied not only the conceptual framework of general systems theory to understanding social interaction, but also used this approach to mathematically model the marital interactions of couples using differential equations. They have been able to predict with 90% accuracy which couples will divorce or stay married, as well as the level of marital satisfaction of those who do stay married, based on the coding of the positive and negative emotions experienced by marital couples during relatively brief discussions of problems within the marriage or other topics. In comparison, understanding marital satisfaction through what are essentially linear approaches (e.g., using self-report questionnaires) has been much less successful (Gottman, 1994, 1999; Gottman et al., 2002).

In many areas of psychology, like in other areas of science, it is not possible to model complex systemic or adaptive change processes using linear conceptualizations of change. Therefore, no comparison can be made between the effectiveness of linear and nonlinear approaches because linear models cannot be meaningfully applied. In cases where it was possible to directly compare the effectiveness of linear and nonlinear models for explaining particular phenomena, Guastello (1995, 2002) calculated the cumulative advantage of the nonlinear over the linear models to be approximately

2 to 1—"That is to say that about 50% of the explanation for a phenomenon comes from knowing what dynamics are involved" (Guastello & Liebovitch, 2009, p. 4). A doubling of the amount of variance explained is remarkable, suggesting that far more complete explanations of psychological phenomena may be possible through the use of nonlinear dynamical systems and complexity theory.

Conclusions

The scientific study of human psychology is a highly challenging endeavor given the tremendous, bewildering complexity of the phenomena involved. Though the development of psychology as a discipline has been complicated and often quite conflictual, the field continues to advance steadily. Indeed, recent progress in psychology and related fields has been fairly dramatic. The findings from these fields have now accumulated to the point where it is time to reassess the theoretical and empirical foundations upon which professional psychology is based.

It was noted in the previous chapter that a conceptual framework that could unify professional psychology as a whole would need to recognize the tremendously complex nature of human psychology as well as accommodate the full range of empirically supported treatments and other psychological interventions (e.g., tests, prevention strategies). It would also need to be applicable across all of professional psychology, including the full range of populations and disorders treated within the field and all the various specializations included within behavioral health care. It would also need to accommodate the already expansive and steadily growing body of scientific knowledge regarding the functioning of the human mind and brain. Most of the traditional theoretical orientations in professional psychology were developed before the most recent period of dramatic advances in the sciences and were not revised to incorporate the findings of this research. In recent decades, the scientific understanding of complex living systems has advanced dramatically, and general systems and nonlinear dynamical systems theory have played critical roles in advancing this understanding. Therefore, these approaches need to be integrated into the theoretical frameworks used to conceptualize human development, functioning, and behavior change.

There is widespread consensus at this point that the complexity of human psychology can be understood only through a comprehensive integrative perspective that captures influences from across several levels of natural organization, from the biological, evolutionary, and genetic to the individual and on through various sociocultural levels. As noted in earlier chapters, the traditional theoretical orientations in professional psychology have often been criticized for giving insufficient attention to the influence of sociocultural factors such as ethnicity, culture, gender, religion, and class (e.g., Sue & Sue, 1990, 2008). Many others have not integrated the important findings from biology, the neurosciences, and evolutionary psychology regarding the highly proscribed anatomy and physiology of the brain and the corresponding implications for behavior (e.g., Buss, 1991; Confer, Easton,

Fleishman, Goetz, Lewis, Perilloux, & Buss, 2010; Pinker, 2002). In addition, the element of time must be incorporated into any conceptualization of human psychology because cognition, emotion, and behavior cannot be understood without integrating a developmental perspective.

There is no debate within the sciences regarding the need to take a systemic view that integrates the interactions between the relevant levels of hierarchical natural organization and across time when attempting to understand human psychology. There appears to be unanimity across the sciences and within psychology that human development and functioning, for either individuals or populations, cannot be understood without a comprehensive perspective on the interactions between biological, psychological, and sociocultural processes (e.g., Capra, 1996; Dawkins, 1976; Engel, 1977; Eysenck, 1997; Wilson, 1998). Attempting to understand important psychological phenomena such as intelligence, personality, or psychopathology through frameworks that omit any of these perspectives will result in seriously incomplete explanations. Even human evolution itself cannot be understood without this type of comprehensive, integrative perspective. For example, in their groundbreaking work, Lumsden and Wilson (1981, p. 1) described "gene-culture coevolution" as "a complicated, fascinating interaction in which culture is generated and shaped by biological imperatives while biological traits are simultaneously altered by genetic evolution in response to cultural innovation."

A metatheoretical perspective is needed to integrate the complexity of these many factors, and the biopsychosocial approach is the obvious candidate for capturing this perspective. There are other concepts or terms that could be used to communicate this type of systemic perspective (e.g., ecological, ecosystemic), but the biopsychosocial term is well known across the health care and social science fields in the United States as well as internationally. Both the biopsychosocial approach and general systems theory on which it is based convey the importance of a holistic systemic approach to understanding human psychology and the importance of information, feedback, and communication in the functioning of open living systems (of which the human being is the prime example). Together with the focus on change emphasized by nonlinear dynamical systems approaches, these perspectives capture quintessential characteristics of human development and functioning. The term *complexity theory* captures this combination well because it incorporates both the systemic and the nonlinear change perspectives. For purposes of efficiency, the term *biopsychosocial approach* will be used in the remaining chapters of this volume to refer to both the biopsychosocial application of general systems theory as well as the associated complexity theory perspective that is used throughout the sciences to understand complex adaptive systems.

An argument can be made that professional psychology currently has no choice but to update the scientific foundations of the field to incorporate a systemic biopsychosocial and nonlinear dynamical systems approach to understanding complex human phenomena. While the linear model is impressively robust and broadly applicable, modern science has found that complex phenomena cannot be understood without using nonlinear dynamical systems concepts. If the foundations of the field are not updated to incorporate these findings, the claim that professional psychology is a science-based profession is called into question. There is a risk to the legitimacy of the field if it falls behind the underlying science.

Updating the scientific foundations of the field needs to be done carefully and responsibly, however. Though the success of modern science is very impressive, it is also critical to appreciate the distance it still has to go to describe and predict the behavior of complex adaptive systems. For example, it is possible that the West, Brown, and Enquist (1999) metabolic scaling theory will be able to unify all of biology under one theoretical framework—a development that would represent a truly monumental achievement. On an even grander scale, perhaps Wolfram's (2002) cellular automata theory will be able to explain complexity at *all* levels of natural organization. These are new proposals, however. While the current evidence supporting their validity is intriguing, much more research is needed before conclusions will be possible. This is typical in science—it took more than six decades for Newton's dynamics to be widely accepted, and other important scientific advances similarly found slow acceptance (Mitchell, 2009).

Though science obviously has a long way to go in explaining nature's complexity, it is important for professional psychologists to note that science began to move beyond the linear conceptualization of natural phenomena a half-century ago already (Stewart, 1989). Systemic and nonlinear approaches have proven essential for understanding and predicting the behavior of complex systems across fields, and human beings are the quintessential complex system (Koopmans, 2009). As a clinical profession based on science, professional psychology needs to incorporate this research.

At the same time, it is important to be cautious and not overstate what is currently known regarding human psychology. The success of nonlinear dynamical systems and complexity theory approaches across the sciences is impressive, and their intuitive appeal is seductive. In many ways, they match our personal experience regarding the tremendous complexity of human experience: the complex interaction of genetic inheritance and experience; the subtle but powerful influence of families, neighborhoods, society, and culture on development and functioning; the difficulty of behavior change in some cases and the ease of behavior change in others; the seductive power of intertia; as well as the big difference that even a very small change (or luck) can cause when circumstances are right.

The understanding of human nature is still quite limited. Though the amount that is known is impressive, many of the tremendously complex interactions between all the intricate brain structures that mediate cognition, emotion, and behavior are not well understood, nor are the complicated interactions between biological, psychological, and sociocultural processes across time. There are major risks associated with prematurely drawing conclusions about the nature of human psychology before the complexity of the human mind and brain is sufficiently well understood. As a science-based profession, professional psychology simply requires adequate scientific verification before hypotheses are accepted as fact and theory.

The biopsychosocial framework as a metatheoretical perspective helps protect against the risk that conclusions about the precise nature of psychological development and functioning will be drawn before sufficient evidence has accumulated.

This framework provides a counterbalance against premature claims that particular biological, psychological, or sociocultural factors are responsible for developmental or behavioral outcomes. While many psychological processes at lower levels of natural organization such as sensation, perception, and learning are already thoroughly understood, less is known about more complex processes at middle levels of complexity such as motivation, memory, emotion, and intelligence. At higher levels of organization and complexity, processes such as personality and psychopathology will be even more difficult to explain and predict, and social behavior can be still even more difficult to explain. This complexity is all incorporated into the biopsychosocial perspective which decreases the likelihood that simplified, reductionistic or unverified claims about the nature of psychological functioning or behavioral change processes are adopted.

The evidence supporting the general framework of the biopsychosocial approach for explaining human psychology at the metatheoretical level is very strong. Indeed, it appears that there is no significant disagreement about the appropriateness of this framework for the field. The hierarchical interconnectedness of the biological, psychological, and sociocultural subsystems and the nonlinear nature of change in living systems is simply incontrovertible. Outmoded conceptualizations of human psychology that do not incorporate these perspectives need to be replaced so that thoroughly established findings regarding the nature of complex living systems are incorporated into the scientific foundations of professional psychology. While fully acknowledging the vast amount of research that still must be done, it is evident that the turning point for the field to adopt a unified science-based conceptual framework that incorporates these perspectives has been reached. The preparadigmatic era of the field has ended. The assortment of theoretical orientations that have historically been used to conceptualize clinical practice in psychology now needs to be replaced by a unified biopsychosocial metatheoretical approach.

6 Ethical Foundations of Professional Psychology

The central role of science in professional psychology is obvious—any modern conceptualization of professional psychology as a health care profession requires that psychological intervention be based on the scientific understanding of human psychology. The central role of ethics in professional psychology is perhaps less obvious. At a basic level, however, the reason for its central role is completely straightforward. While the science of psychology helps explain the nature of personality, psychopathology, and behavior change, it is often not particularly informative with regard to how to apply this knowledge in professional practice. The same is true, of course, regarding the relationship of science and practice in many areas of life—science often cannot provide guidance regarding the surfeit of human decisions that must be made across the many domains of a person's existence. In the case of behavioral health, professional ethics are essential for judging how to apply scientific knowledge within the socially, economically, and legally embedded context of behavioral health care, both at the level of the individual therapist working with individual patients as well as at organizational, institutional, and societal levels.

There are also several specific reasons why professional ethics are centrally important in professional psychology. At a fundamental level, health care must emphasize ethics because of the nature of the responsibilities and relationships involved. It is the role of *care* in health care that causes ethical conduct and moral virtues to be of central importance (Beauchamp & Childress, 2009). Without these, all forms of caring relationships (e.g., parental caregiving of children, elder care, medical and nursing care, mental health care) become seriously diminished— caring, compassion, trustworthiness, integrity, and other moral virtues are fundamentally important in all these relationships. One can argue that these virtues are especially important in psychotherapy due to the critical role of the therapeutic relationship and the vulnerability that patients feel when revealing and processing highly personal and private issues. Behavioral health issues frequently also carry great stigma in society, the effects of which can even outweigh the impairments related to having a mental illness (Hinshaw & Stier, 2008). Therefore, the importance of basic trust and moral virtues in the therapy relationship is very high. Psychotherapists consequently need to aspire to the highest levels of ethical conduct and not merely observe only minimal standards of moral behavior (e.g., merely "follow the law" but not aspire to higher goals).

Foundations of Professional Psychology. DOI: 10.1016/B978-0-12-385079-9.00006-0

The collective wisdom of the many psychologists who helped develop the profession over the years has also strongly emphasized the role of ethics in the field. The current mission statement of the APA states that "The mission of the APA is to advance the creation, communication and application of psychological knowledge to benefit society and improve people's lives" (APA, 2010). It then goes on to list five basic purposes of the organization, two of which focus on ethics, including "the establishment and maintenance of the highest standards of professional ethics and conduct of the members of the Association" (APA, 2010). The APA Ethics Code (2002) is also explicit about the critical role of ethics in the professional behavior of psychologists. It states that "Membership in the APA commits members and student affiliates to comply with the standards of the APA Ethics Code" (p. 407), and "In the process of making decisions regarding their professional behavior, psychologists must consider this Ethics Code in addition to applicable laws and psychology board regulations" (p. 408). These statements give high priority to the role of ethics in the professional lives of psychologists. They also show that the field recognizes the substantial influence and power that psychologists exercise at several levels, from the interpersonal in the therapy relationship, to their involvement in health care, education, and legal systems, and in terms of social policy. Psychologists, both individually and collectively, must be committed to the highest ethical standards when exercising these various types of influence and power in people's lives.

At a practical level, solid knowledge of professional ethics and related legal issues is also required to complete one's graduate education and become licensed to practice in the field. Ethics and legal issues are strongly emphasized on graduate comprehensive exams and on licensure exams. In addition to the heavy emphasis on ethics and legal issues on the Examination for Professional Practice in Psychology (the national licensure exam required in almost all jurisdictions in the United States and Canada; Association for State and Provincial Psychology Boards, 2011), most states additionally require that psychologists pass a state-level exam emphasizing ethical and legal issues. Solid knowledge and application of ethics is also required for students to succeed in their practica, internships, and postdoctoral supervised experience. Ethical behavior is a high priority for employers as well. Experience suggests that one of the quickest ways to enter a formal disciplinary process as a graduate student or an employee in the behavioral health care field is to engage in unethical behavior. Several other types of problems (e.g., an obnoxious personality, poor work habits) are far more likely to be tolerated than unethical behavior.

Many of the issues that patients deal with in therapy also have critically important ethical dimensions. These occasionally involve life and death issues (e.g., suicide) along with many others that are highly consequential in patients' lives (e.g., weighing the impact of abusiveness or boredom in one's career or marriage and what to do about it; decisions about how best to deal with family and parenting problems; dealing with unplanned pregnancy, sexual orientation, or the role of religion or culture in one's life; accepting the consequences of having a mental illness, substance dependence, or personality pathology). Therapists also need to be skilled at handling ethical and legal issues that arise in everyday practice (e.g., how to handle a secondhand report of child abuse, how to respond to a patient who is engaged in behavior that is harming his or her family members without their knowledge, how to respond when one's neighbor is the next patient in the clinic waiting room). Therapists need firm grounding in ethics to be able to ensure that they are acting in patients' best interests and are weighing all the relevant considerations when responding in these situations. This is important for allowing therapists to be confident that they are acting ethically in their own professional practice as well as for helping patients improve their own ethical decision making.

All these considerations point to the central importance of professional ethics in the practice of psychology. From the level of managing the daily responsibilities associated with clinical practice to the fundamental nature of health care, ethics play a central role in professional psychology education and practice. Though the scientific understanding of human psychology steadily advances, science has limited ability to address many questions regarding the appropriate application of behavioral health interventions in people's lives. Empirically derived ethical principles are not yet available, and perhaps never will be. Therefore, a description of the theoretical foundations of professional psychology will be seriously incomplete without an examination of the central role that ethics play in the field.

The Importance of Foundational Ethical Principles

In addition to viewing ethics as foundational to the practice of psychology, this volume takes the position that a grounding in the *foundations* of professional ethics is necessary to gain a deeper, more thorough, and more useful understanding of the role of ethics in professional practice. Of course, it is critical to be knowledgeable about the many specific ethics codes, laws, rules, and policies that govern mental health practice. But having familiarity with only these codes, rules, and policies, without an understanding of the foundational principles from which they are (or should be) derived, can result in an incomplete and superficial analysis of many ethical dilemmas that arise in professional practice.

Being ethical—doing what is right and not doing what is wrong—often appears to be straightforward. For example, following the "Golden Rule" (how would one like to be treated in the same situation) often provides reliable guidance even in difficult circumstances. Upon closer examination, however, it is evident that ethical questions are often much more complex than those that can be adequately addressed by following the Golden Rule. This is particularly true when working in today's increasingly diverse society and interconnected global community which are undergoing rapid technical, scientific, and social change. The following three issues emphasize the complexity of moral issues and the need to take a careful and thorough approach to professional ethics.

Confusing What Is for What Ought to Be

There have been many social practices and conventions that have been thoroughly accepted in one historical or cultural context but are viewed as unacceptable in another time or context. For example, after observing the relationship between those with more and less power in society, it was evident to Herbert Spencer, the father of "Social Darwinism," that the notion of the "survival of the fittest" applied to human social and economic relations just as it applied to the evolution of animal species. He went on to argue that it would be wrong for government to interfere with nature's tendency to let the strong dominate the weak. Spencer failed to note that just because it was true that the powerful tended to dominate the less powerful, that does not imply that this is the way things ought to be (Moore, 1903/1959). There are, of course, many other examples of unjust social practices that were accepted as morally correct largely because they were viewed as common and "natural." A cruel example from the mid-eighteenth century involves a Louisiana physician who identified a new psychiatric diagnosis, *drapetomania*, a running-away-from-home disorder, to diagnose the pathology of slaves who wished to run away from their masters. The recommended treatment for this disorder was a beating (Szasz, 1971).

Recent neuroscience research helps clarify some of the underlying processes involved when humans associate what is common or natural with what is morally correct. For example, research participants were asked to imagine the following two scenarios described by Greene (2003):

• As you are driving through the countryside, a man on the side of the road with serious injuries pleads for you to stop. You pull over and the man explains that he had a hiking accident and needs to be taken to the nearby hospital. You want to help, but if you give him a ride, the blood and dirt may stain the leather upholstery in your car. Is it appropriate for you to leave this man on the side of the road in order to keep your upholstery clean?
• You are opening your mail and read a letter from a reputable aid organization asking for $200 to allow them to provide badly needed medical care to people in a poor part of the world. Is it morally acceptable to pass on this opportunity to make a donation to this organization?

With regard to the first situation, a large majority of people say that it would be horribly selfish to refuse the hiker's request for help (Greene, 2003). The man badly needs medical attention and you are in a position to provide it without great cost. Not to do so would be morally deficient. On the other hand, in the second situation, most people would say it is not morally wrong to ignore the request from the aid organization. It would be admirable to help people in faraway places who have life-threatening medical needs, but we are not obligated to provide assistance. The costs involved in these two situations might be similar (i.e., the cost of getting the car upholstery cleaned may be about the same as the $200 donation). Nonetheless, people tend to respond in opposite ways even though only one person in the first situation would be helped while a much larger number of people would be helped in the second.

Functional magnetic resonance imaging studies conducted by Greene and others (e.g., Greene, 2003; Green, Nystrom, Engell, Darley, & Cohen, 2004; Hauser, 2006) have found that moral dilemmas such as the first one above, which are personal in nature, are associated with greater neural activity in the emotion and social cognition areas of the brain. Research participants experienced strong, immediate feelings regarding the appropriate moral response in that type of situation. On the other hand, the research participants found the second dilemma to be impersonal. This dilemma was associated with greater activity in the cortex regions associated with abstract reasoning and cognitive control, while there was less activity in the emotion and social processing areas and no automatic feelings about the morality of the situation. This finding might be understandable from an evolutionary standpoint because, over the course of human history, an evolutionary advantage may have been gained by those who were concerned about the well-being of those who are close to them. In fact, altruism may have developed as a human characteristic because it was important to the survival of the group and even one's own survival that those who became sick or injured received assistance from the other members of the community. Helping those from outside the community who were sick or injured, however, may have been irrelevant to the survival chances of one's own community, or it could actually even decrease one's chances of survival because outsiders who recover from casualties in warfare, for example, may attack one's community again. The tendency to have strong feelings regarding the needs of members of one's own community, while feeling little concern about outsiders, may have conferred a significant survival advantage. This distinction may be irrelevant, however, when examined from an objective moral perspective where place of residence is not a relevant consideration for judging the value of human life.

There are a variety of psychological responses like the ones described above that involve automatic neural processing that may have conferred an evolutionary advantage at one point but that are not necessarily based on logical or moral considerations (Greene, 2003; Green et al., 2004; Hauser, 2006). For example, if one happens upon a group of delinquent teenagers kicking and beating a cat or dog to death for amusement, most people would immediately judge that behavior to be highly immoral. If one happened upon that event on the way to a restaurant for a steak dinner, however, questions regarding the morality of killing and eating animals may not even enter one's consciousness (Pollan, 2006). Many social situations, such as judging whether a person is beautiful or a man or woman, are also typically processed automatically on an intuitive, emotional basis. The feelings one has about these situations are also typically subjectively felt to be quite natural and correct. Just because the resulting judgments feel so "right" or so "wrong," however, does not necessarily make them correct from an objective moral perspective (see also Ariely, 2009).

The Universality of Ethics

A second complex issue that is very important when working with individuals of diverse ethnic, cultural, and religious backgrounds concerns the universality of morality. For example, to what extent should we accommodate the diversity of viewpoints that exist across cultures and religions regarding the treatment of women, children, and animals? Do our ethics codes and moral standards apply universally, or are moral standards the products of historical and cultural circumstances and so their applicability depends on the specific circumstances involved in any given situation? That is, are ethics basically relative? This question is rapidly growing in importance as societies become more diverse and the global community becomes more interrelated.

There is not complete agreement among ethicists on this question, but certainly many authorities argue that all individuals who are committed to morality, across cultures, time, and place, agree on the basic foundations of ethics, and there consequently does exist a universal common morality (e.g., Beauchamp & Childress, 2009; Council for the Parliament of the World's Religions, 1993; Gert, Culver, & Clouser, 2006; United Nations, 1948). There are many culturally specific aspects of the universal morality that can vary greatly across religious groups, institutions, and even professions, but there is also significant agreement among ethicists that a shared, basic common morality does exist.

Even leaders of the world's religions, who obviously hold widely divergent views on many subjects, have agreed regarding the existence of a basic universal morality. In 1993, the Council of the Parliament of the World's Religions endorsed the following statement in its "Declaration Toward a Global Ethic":

> We are persons who have committed ourselves to the precepts and practices of the world's religions. We confirm that there is already a consensus among religions which can be the basis for a global ethic—a minimal fundamental consensus concerning binding values, irrevocable standards, and fundamental moral attitudes. (p. 3)

A similar effort was undertaken by psychologists as well. In 2002, the International Union of Psychological Science (2005) approved the development of a "Universal Declaration of Ethical Principles for Psychologists." It explicated a general "common moral framework that guides and inspires psychologists worldwide" (p. 2), though it is also "written in language that is generic rather than prescriptive. It deliberately avoids prescribing specific behaviors or standards of conduct inasmuch as these must be relevant to local culture, customs, beliefs, and laws" (p. 2).

Despite the widely diverging viewpoints between many cultures and religions on many issues, there does appear to be widespread agreement that a universal basis for fundamental ethical principles does indeed exist. Of course, applying that universal morality in diverse contexts and cases remains complex, but at least there appears to be a solid starting point for approaching these difficult questions.

The Question of Moral Status

Another critical question central to many ethical dilemmas concerns the moral status of different groups of individuals. Across history, enemies, slaves, animals, women, children, and psychiatric patients have often had lower moral status. Their interests and rights often received lower protection and, in some cases, no protection at all (Beauchamp & Childress, 2009; Lindsay, 2005). This situation has been changing dramatically, however, in recent decades. Though some aspects of ethical theory have evolved relatively slowly over the centuries, more recently there has been a clear increase in the range of individuals to whom moral norms are applied. So while it might appear that societies and institutions have embraced higher moral standards in recent decades and centuries, the bigger difference is that larger numbers of groups of individuals have been extended moral status so that they too are covered by ethical standards that have remained relatively stable over time.

There remain many fascinating questions about whether comatose patients, anencephalic babies, fetuses, embryos, human eggs, and even animals deserve full moral rights (Beauchamp, Walters, Kahn, & Mastroianni, 2008; Jecker, Jonsen, & Pearlman, 2007). In addition, mental health care and medicine often involve caring for children and adults who are ill, impaired, disabled, incompetent, or otherwise vulnerable, and important decisions involving ethical issues often must be made when these individuals are not able to participate fully in the decision making.

This overview of some of the complicated aspects of morality is certainly not intended to resolve questions regarding the appropriate application of ethics in mental health practice. Instead, it is intended to emphasize the importance of a deeper understanding of professional ethics that goes beyond just familiarity with ethics codes and mental health law. It is also intended to point to the need for a biopsychosocial approach to understanding ethics. Even just the brief review above illustrates that an examination of ethics that does not incorporate psychological, sociocultural, and even biological and evolutionary perspectives will be seriously incomplete.

The remainder of this chapter provides an overview of the ethical foundations of the codes, policies, and laws that govern health care in the United States. It is intended to show that the ethical foundations underlying professional psychology as a health care specialization are very well developed, and that it is critical that these foundations be integrated into the comprehensive framework for understanding the basic nature and purpose of the field.

Ethical Theory

Part of the reason that professions develop codes of ethics is self-serving. Ethics codes communicate to the public, governments, and the legal system what members of a profession are allowed and prohibited from doing. These codes then help to prevent the government or others from holding members of the profession accountable for taking certain actions or refusing to take others (e.g., breaking a patient's confidentiality when he or she threatens another person; refusing to release patient information to an inquiring police officer without appropriate authorization). Requiring a commitment from members of a profession to observe ethics codes is also an attempt by the profession to be self-regulating so that the government does not set policies and make basic decisions for the profession.

Codes of ethics are meant to be far more than just self-serving, however. They are based on ethics, a branch of philosophy that is, of course, a very old and well-developed field of scholarly inquiry. The field of ethics is concerned with general ethical theories that provide an integrated body of moral principles for addressing ethical behavior comprehensively. The subfield of *biomedical ethics* is the primary concern of the present chapter, however. This term is commonly used to refer to the ethical principles that guide health care. These principles derive from general comprehensive ethical theories but are applied specifically to health care and biomedical research.

Before examining biomedical ethical principles more specifically, the four general ethical theories that have been most influential on the development of biomedical ethics (Beauchamp & Childress, 2009) will be briefly reviewed. The following overview of these theories is very brief but is intended to convey a sense of their strengths, weaknesses, and importance for informing contemporary biomedical ethics.

Consequentialist Approaches

John Stuart Mill (1806–1873) developed the most influential consequentialist approach to ethics, called *utilitarianism* because of the priority it gives to the principle of utility, which justifies all other principles and rules. Actions are right or wrong according to their balance of good and bad consequences. This approach is often associated with the maxim that "We ought to promote the greatest good for the greatest number," or at least the least disvalue when all the options are undesirable. From this perspective, the ends justify the means if the benefits associated with the consequences of an action outweigh the harms resulting from those actions. Mill and Bentham are considered *hedonistic* consequentialists because they emphasized happiness or pleasure as the goals to be maximized. More recent utilitarians argue that other values such as knowledge, health, success, and deep personal relationships also contribute to individuals' well-being (Beauchamp & Childress, 2009; Griffin, 1986).

Utilitarianism is not a fully satisfactory theory of ethics, however (Beauchamp & Childress, 2009; Cohen & Cohen, 1999; Freeman, 2000). For example, some preferences might be considered to be immoral, regardless of any weighing of harms and benefits (e.g., sadism, pedophilia, inflicting pain on animals). This approach also does not answer the question of whether maximizing value is an obligation that must be observed (e.g., is one obligated to donate one of his or her kidneys because one can be fully healthy without it?). Another problem left unresolved by this approach concerns whether the interests of the majority can override the rights of minorities. Are there cases where the rights and interests of even the smallest minority need to be protected, independent of the weighing of costs and benefits? Should education, police protection, and health care be provided to all individuals in a society, even if it is relatively costly to provide these services to particular groups?

Deontological or Kantian Approaches

Immanuel Kant (1724–1804) emphasized a very different set of obligations than consequentialist theorists. From his perspective, duties (*deon* is Greek for duty), obligations, and rights are the highest authority, and right actions are not determined solely by the consequences of actions. Ends do not justify means if they violate basic obligations and rights, and human beings must always be treated as ends and never as means only (Beauchamp & Childress, 2009; Donagan, 1977). To find a source of ultimate obligations and rights, religious traditions have appealed to divine revelation (e.g., the Ten Commandments) and others to natural law (e.g., "natural law" at the Nuremberg trials). From Kant's perspective, the *categorical imperative* is the highest authority in determining the morality of ethical principles: "I ought never to act except in such a way that I can also will that my maxim become a universal law" (Kant, 1785/1964, p. 96). In other words, morally acceptable decisions are applicable universally, in all situations that are similar in relevant ways.

This approach also has weaknesses. Kant has been criticized for emphasizing reason above all other considerations, including emotion, suffering, and pain, and consequently his arguments against suicide and other issues have been viewed as inadequate (Cohen & Cohen, 1999). This approach also does not resolve situations where individuals have two competing obligations (e.g., taking one's children on a promised trip vs. staying with one's mother who has developed a serious illness). In addition, many of the moral obligations we feel are based on the nature of the relationships we have with family, friends, coworkers, and neighbors—the correctness of our behavior in these situations is significantly affected by the commitments we have to these individuals, not by objective moral obligations to people in general (Beauchamp & Childress, 2009).

Liberal Individualism

Both considering our moral obligations and duties and seeking the greatest good for the greatest number have been very useful in developing functional systems of morality. In the Western world in general and the United States in particular, however, there has also been a strong emphasis on protecting individual rights. Philosophers such as John Locke (1632–1704) and Thomas Hobbes (1588–1679) emphasized the importance of human rights and civil liberties, and their rights-based theorizing has become strongly integrated into the Anglo-American legal system (Beauchamp & Childress, 2009; Dworkin, 1977). Hobbes famously remarked that without strong government that provides basic protection of individual rights, security, and rule of law, life is "nasty, brutish, and short" (1651/2002, p. xiii).

From this perspective, basic human rights to autonomy, privacy, property, free speech, and worship are foundational to the functioning of civil society. Even though these rights are very strong, however, they are not absolute. For example, one's right to life is perhaps the strongest of the rights an individual can hold, and yet it too can be overridden in cases of war and self-defense (whether an exception also applies to those who have been found guilty of capital crimes remains a controversial issue). Rights are considered to be *prima facie* binding; that is, when first considering an ethical situation, rights need to be observed, though they may be overridden in certain circumstances (Beauchamp & Childress, 2009; Dworkin, 1977). In health care, patients' rights to informed consent, the refusal of treatment, lifesaving emergency medical care, and confidentiality all function in this way. Rights also function similarly to obligations. If someone has a right to something (e.g., education, health care), then others have obligations to provide the goods or services needed to provide that right, or to refrain from actions that would violate that right (e.g., not disclose confidential information without an appropriate authorization from the individual). Tensions between having a right to something and the corresponding obligations by others to help provide for that right are a perennial source of conflict between those who would emphasize liberal individualism (i.e., freedom from government intrusions) over those who would emphasize government services and controls to provide for an orderly, secure, and efficient economy and society (e.g., those who emphasize government regulation of industry, education for all children, health care for all citizens).

While there is no doubt that the protection of human rights is critical to the effective and humane functioning of governments and society, rights are also viewed as providing a limited perspective on morality (Beauchamp & Childress, 2009; Cohen & Cohen, 1999). Rights typically do not account for the moral significance of a person's motives, conscientiousness, or integrity. It is not always moral to do what we have a right to do (e.g., if the free market is allowed to determine all prices, can health care providers charge exorbitant rates when someone has a medical or psychiatric emergency and might agree to pay almost any price for the health care that is needed?). Liberal individualism also focuses a great deal on protecting individual rights from government intrusion, while the rights and interests of the community as a whole receive less attention. There is a great deal about community life that has substantial benefit, such as public health, educated citizens, the protection of animals, and culture in general (as Oliver Wendell Holmes remarked, "I like to pay taxes. With them I buy civilization"; 1939, pp. 42–43). In addition, rights-based systems tend to take adversarial approaches to resolving conflicts when one's rights have been violated. There are aspects of family life, for example, where such an adversarial approach may not serve psychologically healthy purposes. For instance, though children have a right to be free from maltreatment, should a parent be held responsible for monetary damages caused by his or her abuse or neglect of a child? When a couple cannot agree to child custody arrangements after they decide to divorce, is an adversarial rights-based approach to defeating one's ex-spouse in court the best approach to resolving that dispute?

Communitarian Approaches

Community-based approaches, the last of the four most influential ethical theories reviewed here, are often critical of the previous approaches. Communitarian

approaches consider the common good, communal values and goals, and cooperative virtues to be fundamental considerations in ethics. From this perspective, too
great an emphasis on individuals' rights and autonomy can result in an uncaring
society where individuals look out for themselves and have little responsibility for
the well-being of others (Beauchamp & Childress, 2009; Freeman, 2000; Sandel,
2005). This can result in a breakdown of family and civic responsibilities and lead
to marital infidelity, welfare dependency, abandoned children and elderly parents,
and even the disappearance of a meaningful democracy. Carol Gilligan's 1982
book *In a Different Voice* was influential in highlighting disadvantages of too
strong an emphasis on autonomy and individual rights. When she challenged the
Freudian notion of the inferiority of women's moral development, she emphasized
that women's "strong sense of being responsible" (p. 21) to family members and
loved ones, as opposed to a strong commitment to autonomy and impersonal rights,
was a sign of moral strength, not inferiority. Militant communitarian views (e.g.,
communism) can be hostile to individual rights, while moderate communitarianism
emphasizes the importance of communal values such as parenting, teaching,
governing, healing, and caring for those who are less able to care for themselves.
These values are especially important in health care where many patients are
physically and/or psychologically vulnerable, and promoting health among individuals in general is in the interests of the common good of the society (Callahan,
1990).

Communitiarianism has been criticized, however, for presenting what may be
viewed as a misleading dichotomy between individual rights and the common
good—either we protect individual rights and autonomy, or we pursue the welfare
of the community as a whole (Beauchamp & Childress, 2009; Cohen & Cohen,
1999). Such a dichotomy is unnecessary. Individuals are inherently social beings,
and so the common good (e.g., functional families, communities, and government)
is necessary for the individual to thrive. At the same time, the autonomy and rights
of the individual need to be protected against oppressive communities that might
otherwise intrude upon and control the individual. Communal goals and individual autonomy and rights both need to be protected so that the interdependent individual and community can both thrive.

An Integrative Approach

None of the moral theories summarized above adequately resolves all moral conflicts, and consequently none, by itself, provides a satisfactory foundation for biomedical ethics (Beauchamp & Childress, 2009; Rawls, 1999). Each has weaknesses
and strengths, and some serve some purposes better than others. For example, utilitarianism is useful for setting public policy, liberal individualism has played an
important role in establishing legal standards, while deontological and communitarian approaches are useful for guiding many health care practices. There also is no
clear evidence that any of these theories should be discarded because each brings a
valuable perspective that the others lack. There is even neuroscience evidence
that the human brain relies on multiple types of information processing when

faced with different types of moral dilemmas, and these may correspond to the different types of priorities associated with the various ethical theories (Greene et al., 2004).

This situation is similar to that found in psychology. With regard to the traditional theoretical orientations in psychology, each has strengths and weaknesses, some are strong in areas where others are weak, and no one theory is satisfactory as a comprehensive explanation of human psychology. It is only through an integration of the various theoretical perspectives that a more comprehensive understanding of human psychology begins to emerge that is adequate for informing mental health practice.

Leading ethicists such as Beauchamp and Childress (2009) and Rawls (1999) use a combination of deductive and inductive approaches to resolve the problem of developing a coherent system of biomedical ethics that can provide well-justified solutions to the wide range of ethical dilemmas encountered in health care practice and research. These ethicists suggest combining the commonsense moral traditions shared by members of a society (inductive) with ethical principles derived from the above theories to provide structure and coherence (deductive). These are then further clarified and refined through a process called "reflective equilibrium" (Rawls, 1999) where common moral beliefs, moral principles, and theoretical propositions are analyzed and critiqued so that the resulting system becomes increasingly internally consistent and coherent. New scientific, technological, and cultural developments can affect this process such that revised moral beliefs are incorporated into the common morality through an iterative process of analysis and critique. It should be noted that social and physical scientists use a similar combination of inductive approaches (e.g., careful observation and verification to develop hypotheses) and deductive approaches (e.g., tests of theory-derived hypotheses) to make improvements in theoretical explanations of phenomena.

Beauchamp and Childress (1977, 2009) applied this procedure in the case of biomedical ethics and derived four basic ethical principles. Their approach, often referred to as the *four-principles approach* or *principlism*, has become the most influential and accepted approach in biomedical ethics in the United States and much of Europe, and perhaps the world (Gert et al., 1997; Schone-Seifert, 2006). Most ethics texts in professional psychology also rely on the foundational principles advocated by Beauchamp and Childress (e.g., Corey, Corey, & Callahan, 2003; Kitchner, 1984; Koocher & Keith-Spiegel, 2008; Welfel, 2010). Ethical theories occasionally disagree over matters of justification, rationale, and method, but there is a great deal of consensus and convergence on these mid-level principles. Because this approach involves more than merely a top-down deductive process based on inviolable precepts, experience and sound judgment are also very important. In this system, the foundational principles are not relative, though particular decisions and judgments can vary according to circumstances. The combination of induction and deduction also means that the application of the general principles is subject to revision based on the evolution of scientific developments as well as social and cultural practices (Beauchamp & Childress, 2009).

Principle-Based, Common Morality Approach to Biomedical Ethics

The principlism or four-principles approach to biomedical ethics developed by Beauchamp and Childress has been the most influential and widely accepted approach to ethics in health care generally and in psychology specifically (e.g., Corey et al., 2003; Gert et al., 1997; Kitchner, 1984; Koocher & Keith-Spiegel, 2008; Schone-Seifert, 2006; Smith, 2000; Welfel, 2010). It is very important to have an appreciation of these foundational principles so that ethics codes, laws, policies, and rules are not applied in a perfunctory, mechanical manner that is insensitive to the broader ethical considerations that are relevant in particular cases. An analogous situation would be learning a manualized therapy treatment without a full appreciation of its underlying psychological principles and the appropriate situations to which it should be applied. One might learn to implement the treatment in a consistent and reliable manner but may not be able to determine when it should be modified or not implemented at all depending on dual diagnosis, socio-cultural, and other considerations. Therefore, to illustrate how a sophisticated approach to ethical behavioral health practice requires an appreciation of foundational knowledge in biomedical ethics, the four principles in the Beauchamp and Childress approach are briefly reviewed next.

Respect for Autonomy

The word *autonomy* is derived from the Greek words *auto* (meaning self) and *nomos* (meaning rule or governance). Autonomy refers to self-governance of the individual or personal rule of one's self. The concept was originally applied to independent city-states but has since been applied to individuals. To be fully autonomous (e.g., to be completely free from control by others, the source of one's own values, beliefs, and life plans) is unrealistic. Humans are highly social animals, and life in modern democratic societies in particular requires high levels of accommodation, collaboration, and participation. Even factors as personal as one's self-identity, values, and beliefs are shaped by socialization and relationships. Therefore, the focus here is on being substantially autonomous because absolute autonomy is an unrealistic ideal that has limited practical relevance (Beauchamp & Childress, 2009).

Incorporating autonomy into the social order requires not just allowing individuals to claim a right to autonomy, but also a basic respect of others as autonomous beings (Beauchamp & Childress, 2009). For example, if a woman or ethnic minority individual hopes to be judged on the basis of merit for a job promotion or admission to graduate school, but those making the promotion or admissions decisions employ bias or favoritism based on group membership, the individual's merit may have no impact. Therefore, the emphasis of this principle is on respecting others' rights to autonomy, not just simply claiming a right to autonomy. Part of the result of this distinction is an emphasis on working to overcome barriers and obstacles that prevent people from acting autonomously. The efforts of the American Civil Liberties Union (ACLU) to protect the free speech rights of groups such as the American Nazi Party represent examples of this principle, even when these groups express views and values with which the ACLU adamantly disagrees (e.g., as when the American Nazi Party planned to hold a parade in Skokie, Illinois, in 1977 and the ACLU defended their right to assembly and free speech when the city, where one in six residents was a survivor or directly related to a survivor of the Nazi Holocaust, attempted to stop the parade; Strum, 1999). This distinction also has important implications for health care because it implies that professionals have an obligation to provide information and foster autonomous decision making on the part of patients. Because of the unequal distribution of knowledge between professionals and patients, professionals are obligated to provide understandable information and explanations and foster voluntary and adequate decision making by patients.

This situation quickly becomes complicated, however, because many individuals are not in a position to act autonomously. As a group, children are not considered to be able to understand and protect their own interests and welfare and consequently are given few of the rights accorded adults. In mental health care, suicidal individuals in crisis or those with cognitive disabilities or impairments may not be able to make decisions in their own best interests. As a result, in certain circumstances psychologists may themselves determine the best interests of these patients and control their behavior in order to protect them from harm.

The principle of respect for autonomy supports many specific ethical rules such as tell the truth, help people make important decisions when asked, respect people's privacy, protect confidential information, and obtain informed consent (Beauchamp & Childress, 2009). The history of informed consent provides a particularly apt illustration of the importance of this principle, and so will be briefly reviewed in more detail.

Informed Consent

The horrifying accounts of medical experimentation in Nazi concentration camps that were disclosed at the Nuremberg trials following World War II resulted in shock and concern among physicians and medical researchers regarding the rights of individuals who participate in medical research. This eventually led to the general informed consent guidelines that are still in use today. These include the recognition that health care providers and researchers need to obtain informed consent to enable autonomous choices by patients and research participants while also minimizing risks of harm. Researchers and clinicians must also conduct a cost−benefit analysis to help ensure that the benefits of research or treatment outweigh the risks of harm caused by the research or treatment.

In addition to the horrible crimes committed in Nazi concentration camps, there have been many other egregious violations of individuals' rights to informed consent over the years. Many dangerous experiments were conducted on prisoners, reform school residents, and other institutionalized individuals (e.g., exposing people to radiation or injecting them with deadly diseases or toxic substances; Loue, 2000; Washington, 2007). The most famous violation of informed consent in the United States involves the syphilis study conducted in Tuskegee, Alabama, from 1932 to 1972 to determine the natural history of untreated latent syphilis (Jones, 1981). The study was fully approved by the US Public Health Service. It included only African American men, with 399 in the experimental group and 201 in the control group. The men in the experimental group had previously contracted syphilis— they were not given syphilis by the researchers. They were lied to for decades, however, about the medical procedures they received and the effective treatments that became available but were intentionally being withheld (e.g., penicillin starting in the 1940s). The researchers met with physicians from the area where the men lived and provided them with the men's names and a directive that the men not be given any treatment for their syphilis. The men were given painful and risky spinal taps that were deceptively called "special treatment" so that the number of syphilis bacilli in their cerebral spinal fluid could be monitored. The researchers even paid for funeral expenses when the men finally died so they were able to conduct autopsies on the bodies to examine the progression of the disease without the knowledge or consent of the patients or their families. Victims of the study included the men who died from syphilis, wives who contracted the disease, and children born with congenital syphilis (Jones, 1981).

The Tuskegee syphilis study is the best-known example where Americans with less social and political power were exploited by medical researchers. Unfortunately, this has happened many times before and since. For example, the forced sterilization of African American women started during slavery but continued until recently. In a 1991 experiment, the long-acting contraceptive Norplant was implanted into uninformed African American teenage girls in Baltimore—a study that was applauded by some community leaders. Another example involved the testing of dangerous experimental AIDS drugs on foster children in New York City from 1988 until 2001. Eighty percent of the children were African American, and parental consent was not obtained in many of the cases (Washington, 2007).

Nonmaleficence

The principle of nonmaleficence is commonly associated with the maxim to "Above all, do no harm." This principle is implied in the Hippocratic Oath and is often considered the fundamental principle of the health care professions. It is a relatively strong obligation that is distinct from beneficence (provide benefit), the principle discussed next. For example, individuals have a very strong obligation not to push someone off a bridge and into a river, but a much weaker obligation to jump in and attempt to rescue someone who has accidentally fallen in. In general, the obligation to not harm individuals is quite strong, while the obligation to benefit and help others is weaker. The obligation to provide benefit is sometimes quite strong, however, as in the case of child-rearing or health care.

While the implications of nonmaleficence for intentional harms are generally obvious, the implications involving unintentional harms are much more complex and subtle. An important implication of this principle for psychotherapists concerns incompetence. Harm can be caused by omission as well as commission, often by imposing risks of harm through either ignorance or carelessness (Beauchamp & Childress, 2009; Sharpe & Faden, 1998; Stromberg et al., 1988). Examples of this would include having insufficient training and supervised experience to competently diagnose particular disorders or offer certain interventions, to complete an adequate suicide risk assessment or treatment plan, or not appropriately managing countertransference. If patients are harmed as a result, the therapist can be judged negligent, which can then be grounds for a finding of malpractice. The critical question at issue in these cases involves determining whether the professional was practicing up to the *standard of care* for the profession. A professional is not expected to practice at an "expert" level, but he or she can be found negligent if his or her practice falls below current professional standards for competent practice (Koocher & Keith-Spiegel, 2008; Stromberg et al., 1988). Therefore, it is critical that psychologists obtain the appropriate levels of training and supervised experience to be able to competently conduct the assessments and interventions they offer to particular patient populations. When preparing to enter one of the specialty areas within the field, the training and clinical experience required is deeper though narrower compared to the broader experience required for general practice. It is also important for psychologists to maintain their competence and keep up with current standards of practice. This is the rationale for the continuing education requirements that exist across the country for psychology licensure.

Concern regarding the safety of medical interventions in the United States grew dramatically following the publication of the Institute of Medicine report *To Err is Human* in 2000. This report famously estimated that 44,000—98,000 Americans die each year as a result of medical errors—"a jumbo jet a day"—involving misdiagnosis, medications, infections acquired while receiving health care, and wrong-site surgery. This report stimulated the development of the modern patient safety movement that created a variety of pressures to improve patient safety. In the years following the report, accreditation and licensing became more rigorous (e.g., JCAHO began unannounced hospital surveys, duty hour limits were established for medical residents, and most US states mandated the reporting of serious adverse events; Wachter, 2009). More recently, attention has turned to the impact of diagnostic errors. In the first large physician survey regarding this issue, Schiff et al. (2009) found that practicing physicians readily recalled and volunteered information regarding missed or delayed diagnoses. A wide range of diagnoses were missed, including depression that led to a suicide attempt. Autopsy studies have also found that diagnostic errors are frequent and contributed to patient death in approximately 1 in 10 cases (Wachter, 2009).

The safety of psychotherapeutic interventions has also long been a concern; in fact, Freud's concern about his role in potentially causing harm in the case of *Dora* led it to become one of the most famous case studies in the history of psychotherapy. The risks of psychotherapy received relatively little empirical research attention until recently, however. Bergin (1966) began investigating patient deterioration that appeared to be caused by psychotherapy, but little attention was given to the issue until the 1990s when the controversy regarding repressed memories of child abuse grew highly contentious. Other recent therapies for which there is evidence of potential or actual harm include "rebirthing" attachment therapy (Chaffin et al., 2006), group interventions for antisocial youth (Weiss et al., 2005), "conversion therapy" for gay and lesbian patients (American Psychiatric Association, 2000b), critical incident stress debriefing (Mayou, Ehlers, & Hobbs, 2000), and grief therapy (Bonanno & Lilienfeld, 2008). Research clearly also indicates that individual therapists vary greatly in terms of their effectiveness (e.g., Ackerman & Hilsenroth, 2001; Lambert & Barley, 2002; Wampold, 2001), a finding that obligates therapists and their supervisors to ensure that therapist behaviors and practices that are associated with patient deterioration and non improvement are identified and changed. (Practices and procedures that can be used for carrying out these obligations will be discussed in Part III of this book.)

Beneficence

The overarching purpose of psychological practice is to provide benefit to individuals and society. Morality requires that health care providers not only respect patients' autonomy and not harm them, but also contribute to their well-being. The obligation of beneficence includes the provision of benefits (i.e., promoting welfare as well as preventing and removing harms) and the balancing of benefits and harms in an optimal manner (Beauchamp & Childress, 2009). As noted above, the obligation of beneficence is typically weaker than the obligation of nonmaleficence. While sacrifice and altruism are admirable ideals, a person is not morally deficient if he or she does not always provide beneficent acts to others (e.g., donate a kidney to a stranger who needs one). Another implication of the weaker obligation of beneficence is that psychologists typically do not need to take on all patients who are in need of psychotherapy (e.g., they can refer to others those cases where countertransference or other issues are too great a concern).

The principle of beneficence also involves the obligation to optimally balance the risks and benefits of psychological treatment. For example, potential risks of psychotherapy include experiencing strong, aversive, painful feelings and memories, or risks of marital dissolution or loss of employment if a person becomes more assertive or changes his or her life goals. Another important risk involves the consequences of not addressing problems that an individual faces. Without treatment, many problems will not get better or will actually get worse (e.g., emotionally, interpersonally, vocationally, academically, physically, legally). The famous Tarasoff legal case presents a notable example of how benefits and harms need to be balanced. A major potential benefit could have been provided to Tatiana Tarasoff (i.e., her life potentially being saved) if the confidentiality of the patient who represented a threat to her had been broken (i.e., a violation of respect for autonomy and nonmaleficence). The Supreme Court of California judged that the potential harm caused by breaking the patient's confidentiality in order to warn

Tarasoff was outweighed by the potential saving of her life (*Tarasoff v. Board of Regents of the University of California*, 1976).

Even though individual autonomy is highly valued in the United States, Americans commonly accept many limits on their autonomy because it is in the best interests of the individual and the community. For example, traffic laws, air travel security restrictions, drinking water treatments, and medical restrictions are commonly accepted without major questioning. There is generally widespread agreement that strong forms of beneficence, also referred to as paternalism, are justified in order for society to function in a secure, efficient manner (Beauchamp & Childress, 2009). A strong application of paternalism in behavioral health care involves suicide intervention. The question here is whether psychologists are obligated to control suicidal patients' behavior through involuntary hospitalization in order to prevent them from harming themselves. That is, when individuals are unstable, in serious distress, and at risk for causing irreversible harm to themselves through suicide, are professionals obligated to at least temporarily restrict their rights and control their behavior in order to prevent a serious harm from occurring? (Behavioral health and medical patients who decide to end their lives after engaging in careful, logical consideration of the issues may represent a different case, however; Beauchamp et al., 2008.)

Justice

The principle of justice focuses on approaches to fairness and how to ethically distribute the benefits and responsibilities of society. For many purposes, it focuses on how to determine the ways in which we are equal to each other compared to situations when one person has an advantage over another. If there were no limits to resources or opportunities, this would not be a problem. When resources or opportunities are limited, however, these problems can quickly become controversial (e.g., Should prisoners get organ transplants when there is a shortage of organs? Should affirmative action considerations be applied when it comes to making university admissions decisions?).

The minimal principle of justice is often attributed to Aristotle, who argued that equals must be treated equally and unequals must be treated unequally (Beauchamp & Childress, 2009). That is, no person should be treated unequally, despite obvious differences between individuals, until it has been shown that there is a difference between them that is *relevant* to the treatment at stake. For example, mostly everyone would agree that all children should be provided a free public education, despite many obvious differences among children, because those differences are not relevant when judging the value of education. The major problem with Aristotle's approach is that the criteria for judging which differences are relevant are not specified (Rescher, 1966). For example, when individuals need behavioral or physical health care but they are not facing a life-threatening issue, is their ability to purchase treatment relevant to the decision regarding whether or not they will receive services?

Societies usually use a variety of methods for distributing the benefits and responsibilities of communal life (Rescher, 1966). Everyone is given an equal share of some things, such as an elementary and secondary education for all children in many societies. The selection of those who can attend college and graduate school, however, is supposed to be based strictly on merit, as are jobs and promotions. Some benefits of society are decided on the basis of need (e.g., unemployment compensation, social security disability benefits, welfare services), while salaries are generally determined on the basis of the free market. The question regarding whether health care should be provided to everyone, regardless of ability to pay, has been quite controversial in the United States. The question of whether mental health care and substance abuse treatment should be provided has been even more controversial (Cummings, O'Donohue, & Cucciare, 2005).

Many political disputes center around which approach to deciding the distribution of benefits and responsibilities is viewed as fairest. Communitarian approaches tend to emphasize need and commonalities between individuals, while libertarian approaches emphasize liberty and fair procedure. Utilitarian approaches emphasize a mixture of criteria so that the public utility is maximized, the approach usually used in the West. Societies (and even regions within the United States or within individual states) often differ in the emphasis given to these various approaches but normally use several of these principles when developing law and policy (Beauchamp & Childress, 2009; Rescher, 1966).

These four foundational principles provide critical perspectives for developing ethics codes, policies, and laws, as well as deciding how to respond to ethical dilemmas encountered in daily clinical practice. Balancing these various considerations can be complicated, but together they provide a very useful foundation for biomedical ethics. Nonetheless, they are widely considered to be insufficient. One additional perspective is needed to address an important deficiency with these principles.

Moral Character

So far, the discussion of ethical theory has centered on principles, rules, obligations, and rights. These tend to emphasize behavior and the actions one does or does not perform. *Character ethics*, on the other hand, emphasize the *actor* (Beauchamp & Childress, 2009; Cohen & Cohen, 1999; Freeman, 2000). Moral character and virtues have received increasing attention in biomedical ethics because principles, rules, and rights can be impersonal, insensitive, and not always the highest priority in health care and interpersonal relationships. As health care became industrialized in recent years, concern about the level of personal commitment of professionals to their patients has also grown. This too has raised interest in the moral character of health care professionals.

There is significant consistency among characterizations of virtuous health care professionals (e.g., Cohen & Cohen, 1999; MacIntyre, 1982). Beauchamp and Childress (2009) have focused on the following five virtues.

Compassion

Caring and compassion are fundamental to humane health care and are consequently emphasized across health care specializations. This does not imply that health care professionals should be overly or passionately involved with their patients. Too much caring can result in a loss of objectivity and judgment. Instead, Beauchamp and Childress (2009) suggest that an empathic concern mixed with an objective evaluative perspective serves patients' interests most effectively.

Discernment

Discernment refers to the ability to make decisions and judgments without undue influence by extraneous considerations, fears, and personal attachments (Beauchamp & Childress, 2009). Aristotle defined "practical wisdom" as understanding how to act with the right intensity of feeling, in the correct manner, at the right time, and with the proper balance of reason and emotion. For example, some individuals are quite adept at saying just the right thing at the right time, and knowing when not to say anything at all. When a therapy patient is very upset, for example, being able to correctly discern when to provide comfort and reassurance versus when to remain silent and allow the patient to access additional emotions and thoughts is a specific example of this ability in terms of therapy skills.

Trustworthiness

This refers to the confidence that one will act with the right motives and apply the appropriate moral norms when encountering a particular situation. Trust has long played a central role in health care, though distrust became a significant concern more recently as health care in the United States became industrialized and managed-care companies limited or provided incentives to limit care. There is also concern that physicians are practicing "defensively" as a consequence of the increase in malpractice lawsuits in recent decades. Due to the highly personal nature of the psychotherapy process, therapists' trustworthiness is an especially important concern for professional psychology.

Integrity

Conflicts between one's core moral beliefs and the demands of mental health practice can be wrenching. Some strongly held political or religious beliefs can also impair one's ability to work effectively with certain patients. Patience, humility, and tolerance are all critical in mental health care practice, especially in pluralistic, democratic societies and particularly in psychotherapy, where therapists often learn a great deal about patients' personal beliefs, values, and past behavior. But that does not suggest that one must compromise one's values and beliefs—compromising below a certain threshold of integrity means you lose it. Instead, when these types of conflicts arise, psychologists can refer patients to other therapists.

Conscientiousness

Some people are very capable of judging the right course of action in problematic situations, but they are not interested in taking the actions needed to cause the situation to be corrected. Conscientiousness refers to figuring out what is the right response to a situation, intending to carry it out, and exerting the appropriate level of effort to ensure the actions are carried out effectively (Beauchamp & Childress, 2009).

All these virtues fall on a continuum that ranges from ordinary to extraordinary moral standards, from the level of the common morality (that applies to everyone) to the morality of aspiration (Beauchamp & Childress, 2009). While we are all bound to the standards of common morality, we are not bound to more excellent, heroic, and saintly ideals, though we should aspire to them.

Conclusions

The previous two chapters argued that professional psychology needs to be founded on a unified science-based biopsychosocial approach to understanding psychological development and functioning. This chapter argues that scientific knowledge alone is insufficient for appropriately applying that knowledge in clinical practice. Therefore, psychological science and professional ethics are both essential to the safe, effective, and responsible practice of psychology. Firm foundations in both science and ethics are necessary for a comprehensive, unified framework for practicing psychology.

Though biomedical ethics is just a young field, since the 1970s it has quickly become very influential for informing standards, policies, and procedures that guide health care practice and research. Familiarity with this field is critical for engaging in the ethical practice of psychology, and particularly in a diverse, evolving, pluralistic, and democratic society. An appreciation of ethical theory and principles is also critical to discussions regarding how health care, other human services, and social policy can be improved.

The importance of ethics in professional psychology is growing as changes in society, science, and technology are presenting new challenges and opportunities in behavioral health care. For example, teletherapy and the sharing of electronic health care records are now possible as the result of the Internet and other communication technologies. These technologies are highly useful for several purposes but present new concerns about security and one's ability to maintain control of one's privacy. A variety of emerging medical technologies are raising important new ethical challenges as well (e.g., enhanced intellectual performance, genetic testing of embryos, physician-assisted death and euthanasia; Beauchamp et al., 2008;

Jecker et al., 2007). The increasing diversity of society can also introduce conflicts between respecting others' beliefs and traditions and protecting individual rights. There are many topics that are controversial within and across cultures (e.g., divorce, abortion, homosexuality, gender roles, arranged marriages, the control and discipline of children). Harm can easily result from a lack of familiarity with a patient's culture, the ethical and family values generally observed within that culture, the specific beliefs and values of the individual patient and his or her family, and the interaction of these factors with mental health (Knapp & VandeCreek, 2007; Sue & Sue, 2008). As with many of the difficult situations faced by individuals in any culture, these cases often involve a balancing of benefits and harms (e.g., the autonomy of a patient facing an arranged marriage that he or she does not want vs. the alienation from his or her family, religion, and culture that results if the marriage is not accepted). Questions regarding the universality of ethics also grow in importance as societies become more diverse and globalism increases. Knowledge of and commitment to ethical principles and moral character are necessary for finding optimal solutions to all of these challenges and opportunities.

This volume argues that a biopsychosocial approach is necessary for a comprehensive understanding of human psychology. This is true of ethics as well—an integrative approach that incorporates psychological, sociocultural, and even biological perspectives is necessary for a comprehensive understanding of ethics and moral behavior. Failing to integrate multicultural perspectives or findings from neuroscience and evolutionary and developmental psychology and other areas will result in an incomplete understanding of ethics and moral behavior.

The biopsychosocial approach to conceptualizing professional psychology has additional ethical implications due to its emphasis on prevention, optimal health and functioning, and the interrelatedness of psychological, sociocultural, and biological functioning. For example, if prevention strategies are available that can effectively and economically prevent psychopathology, what are the ethical implications of not implementing those strategies? To what extent are psychologists obligated to work to improve the health and functioning of the 80% of Americans who are functioning less than optimally (i.e., are not "flourishing"; see Keyes, 2007, and Chapter 3 in this volume). Given the interrelatedness of psychological, sociocultural, and biological functioning, how much attention should psychologists give to promoting health and functioning across all these domains, as opposed to primarily treating just behavioral health problems? The comprehensive holistic biopsychosocial approach challenges the field to consider incorporating these additional perspectives into the ethical practice of psychology.

7 A Unified Conceptual Framework for Professional Psychology

This volume argues that professional psychology needs a unified science-based framework that will provide a common perspective for conceptualizing education, practice, and research in the field. The lack of such a framework has caused a great deal of conflict, controversy, confusion, and inefficiency for the field. The science of psychology has now progressed, however, to the point where a unified framework for understanding human development, functioning, and behavior change is possible. Adopting such a framework will allow the field to leave behind its pre-paradigmatic past and move ahead with a unified perspective and sense of purpose to more effectively meet the behavioral health and biopsychosocial needs of the public we serve.

The previous chapters examined the various components of a unified conceptual framework for the field. In Chapter 2, the argument was made that professional psychology needs a clearer definition so that the primary purposes and conceptual bases of the field are more explicitly identified. A definition of the field was then proposed, and Chapter 3 illustrated how that definition would clarify the nature of education and practice in the field. Chapters 4 and 5 examined the scientific bases underlying the field and argued that a contemporary scientific perspective on human psychology needs to be based on a biopsychosocial metatheoretical framework in order to represent the highly complex nature of human psychology. It was then argued in Chapter 6 that science alone is insufficient for guiding professional psychology as a health care specialization and that professional ethics need to be fully integrated into the fundamental conceptual foundations of the field.

The present chapter integrates these various considerations into a unified framework for conceptualizing professional psychology and then discusses the main implications of this framework for education and practice in the field. It is argued that a unified paradigmatic conceptual framework for education and practice in professional psychology can resolve many historical tensions that characterized the pre-paradigmatic history of the profession. The proposed framework can be applied across professional psychology as a whole, including all the general and specialized areas of practice. It is based squarely on science and accommodates the wide range of empirically supported treatment approaches that have been used in the field. It also provides a common language and conceptual framework that practitioners, researchers, educators, and students from across subfields and specializations can use to communicate with each other and with those in other health care and human service fields.

Foundations of Professional Psychology. DOI: 10.1016/B978-0-12-385079-9.00007-2

This chapter begins by reviewing the conclusions reached in the previous chapters in this volume and brings those conclusions together under one unified framework. Following that is a discussion of the main implications of this framework for professional psychology education, research, and practice and for resolving theoretical and conceptual inconsistencies in the field.

Underlying Assumptions of the Proposed Unified Framework

The previous chapters argued that the explicit and implicit frameworks used to structure and organize education and practice in professional psychology need to be updated in light of evolving scientific evidence and clinical and educational practices. As part of that analysis, a definition of professional psychology was offered in Chapter 2 to help clarify ambiguities associated with previous definitions of the field. That definition includes the essential components of the unified framework for professional psychology proposed in this volume, and so it provides a good starting point for outlining this framework. It reads as follows:

> *Professional psychology is a field of science and clinical practice that involves the clinical application of scientific knowledge regarding human psychology and professional ethics to address behavioral health needs and promote biopsychosocial functioning. As a health care specialization, it provides psychological services to meet the behavioral health and biopsychosocial needs of the general public. It includes general as well as specialized areas of practice.*

The first sentence of this definition identifies professional psychology as an applied field that is based on science and professional ethics and also notes the main overarching purposes of the field. The second sentence focuses on the clinical role of the field as a health care specialization, the primary purpose of which is to meet the behavioral health and biopsychosocial needs of the general public. The third sentence notes that there are a variety of general and specialized areas of practice within the field, implying that all are united through this common definition of the field. The conclusions underlying this definition include the following:

1. Professional psychology is an applied science.
2. The foundations of professional psychology are scientific knowledge and professional ethics. These provide the fundamental justification and rationale for the practice of psychology.
3. The main purposes of professional psychology science and practice are to understand and treat behavioral health needs and promote biopsychosocial functioning.
4. A primary role and purpose of professional psychology is to provide health care. As a health care specialization, its primary purpose is to meet the behavioral health and biopsychosocial needs of the general public.
5. These conclusions apply across the whole of professional psychology, including all the general and specialized areas of practice encompassed within the field.

These conclusions might seem noncontroversial at first glance. They all seem to be common sense approaches to understanding the role and purposes of the field and the responsibilities of practicing psychologists. They each, however, have important implications for learning and practicing the profession that depart in significant ways from traditional approaches. These will each be considered in turn.

The first conclusion notes that professional psychology is an applied science. As such, science is the authority on which the field rests. This does not imply that non-experimental fields do not help inform psychological practice. Observations and analyses from the humanities and the arts are often critical to the understanding of complex human phenomena. It does focus attention, however, on the validity of evidence and the whole body of knowledge that is available regarding a phenomenon. From this perspective, relying on just the evidence produced within a particular specialization or theoretical orientation would not be sufficient to justify psychological intervention. Indeed, relying on just the evidence produced within all of psychology would be insufficient. The biopsychosocial nature of human psychology requires the integration of knowledge from across disciplines. This level of integration is highly complicated but is nonetheless unavoidable when attempting to understand the tremendous complexity of human psychology.

The second conclusion emphasizes the fundamental importance of psychological science and professional ethics in the practice of psychology. This might seem to be a thoroughly obvious point, one that would be hard to argue against. It seems self-evident that the science of psychology must inform behavioral health care—this has been one of the most deeply held values of professional psychology throughout its history. In addition, due to the nature of psychotherapeutic work with patients within the context of health care, the importance of professional ethics in behavioral health care is also patently clear. Psychologists already have a very strong commitment to both science and ethics.

With regard to the profession's stance on ethics, there is no doubt that psychologists are highly committed to professional ethics. It is also self-evident that science alone is insufficient for guiding clinical practice. As a result, professional ethics are, and must continue to be, fully integrated into clinical training and practice in the field. Given the analysis of the importance of ethics in the profession in the previous chapter, it can also be argued that the field's commitment should be broadened and deepened. There may be a legitimate concern that present approaches to teaching ethics in the field tend to focus on ethics codes, policies, and laws, with insufficient attention given to the foundational ethical theories and principles upon which the codes, policies, and laws are based. A biopsychosocial perspective also needs to be applied to gain a thorough understanding of ethical reasoning and morality. The present volume takes the position that the field's already strong commitment to professional ethics needs to be reaffirmed and even further deepened.

There is also no question that professional psychologists are fully committed to the science of psychology. Indeed, the scientist–practitioner training model that has dominated much of the history of professional psychology gives as much weight to learning scientific methods as it does to clinical practice. The part of that commitment that may need strengthening, however, is the commitment to staying current with scientific advances. The analyses in Chapters 4 and 5 suggest that outmoded theoretical conceptualizations should be replaced and that the field needs to integrate recent scientific approaches to understanding complex living systems. At the metatheoretical level in particular, there is no question that a scientific approach to understanding human psychology requires a comprehensive systemic approach based on the interactions among the biological, psychological, and sociocultural levels of natural organization. The scientific foundations underlying education and practice in professional psychology need to reflect this perspective. So while there is no disagreement that professional psychology is a science-based profession, there is a need to continually update the theoretical and conceptual frameworks used to understand clinical practice.

The fourth conclusion above identifies health care as the primary role and responsibility of professional psychology. An important implication of this conclusion is that psychologists are obligated to care for patients in a manner consistent with health care principles. These principles are somewhat different from those used in service industries where clients are largely responsible for making decisions about the services that they purchase. The obligations surrounding nonmaleficence, beneficence, justice, respect for autonomy, and moral character all tend to be stronger in the health care context than in a service industry. As a result, the importance of employing evidence-based practices, monitoring treatment outcomes, and working collaboratively with other health care professionals to meet patients' needs is greater as well.

Emphasizing the health care role of professional psychology also focuses attention on meeting the behavioral health and biopsychosocial needs of the public. In traditional approaches to learning professional psychology, students often adopted one or more of the traditional theoretical orientations and then based their approach to clinical practice on that orientation. Cases were assessed and conceptualized through the lens of that orientation, and treatment was provided according to the dictates of that approach. Though there was normally an expectation that students would be able to assess and treat a range of issues and work with a variety of different sociocultural populations, the theoretical approach that one adopted had major implications for the interventions one would be competent to offer, the disorders one could competently treat, and the types of patients for whom the treatment would be appropriate.

The approach advocated here is essentially different. In particular, the starting point for learning the profession is different. Instead of focusing on selecting and learning a particular theoretical orientation, the focus is on learning the knowledge and skills required to provide the psychological services that will meet the behavioral health needs of the public. A lack of familiarity with a full range of biological, psychological, and sociocultural influences on individuals' lives will result in a limited perspective on their development and functioning. Therefore, a comprehensive, biopsychosocial perspective on psychology needs to be the basis for learning to meet the needs of diverse members of the public and the many different types of issues they bring with them into treatment.

Previous definitions of professional psychology tended to be unclear about these purposes (see Chapter 2). The present approach is explicit about these purposes, however, emphasizing the health care orientation of the field and working to meet the behavioral health and biopsychosocial needs of the general public. Making these purposes explicit clarifies the role and responsibilities of professional psychologists to their patients and the public. Many aspects of education, practice, prevention, and public policy are clarified by being explicit about the primary purposes of the profession.

The fifth conclusion above emphasizes the importance of this perspective for professional psychology as a whole. The depth of disagreement and conflict between competing theoretical camps and schools of thought in psychology in the past raised serious questions about the validity of the different approaches to practice, education, and research. Indeed, this was the defining characteristic of professional psychology as a pre-paradigmatic field. For a theoretical framework to be scientifically valid for explaining human development and functioning, the psychological community as a whole needs to be convinced by the weight of the accumulated scientific evidence. If certain specializations within the discipline or scientists studying particular aspects of psychology (e.g., neuroscience, multicultural issues) do not accept the validity of a particular theoretical perspective, then that perspective would normally be called into question. For a scientific field to become paradigmatic, its unifying theoretical framework needs to survive this type of scientific scrutiny.

A Unified Framework

The essential components of the definition of professional psychology noted above can now be integrated through a single unified conceptual framework that applies to the entire field as a whole. Essential to this framework is the view of professional psychology as an applied science that is founded on both scientific knowledge and professional ethics. Identifying the primary purposes of the field—namely, to address the behavioral health and biopsychosocial needs of the general public—is also essential to this framework. From the preceding discussion, it is also evident that a biopsychosocial approach is needed to meet these purposes. Given the discussion in the previous chapters, it should be evident that a metatheoretical framework based on an integrative biopsychosocial perspective is necessary for understanding human development, functioning, and behavior change across all the general and specialized areas of psychological practice.

Though the general practice areas of clinical and counseling psychology have not relied extensively on the biopsychosocial perspective, this approach has been endorsed by several of the specializations, and particularly those where the biopsychosocial realms interact in clear and direct ways such as in child, school, health, addiction, neuropsychology, geropsychology, and psychopharmacology (e.g., LeVine, 2007; Martin, Weinberg, & Bealer, 2007; Seagull, 2000; Shah &

Reichman, 2006; Sperry, 2006; Suls & Rothman, 2004; Williams & Evans, 2003). The biopsychosocial approach is implied in the APA accreditation standards which require that all accredited educational programs cover the biological, psychological (i.e., cognitive, affective, and individual differences), and social bases of behavior (APA Commission on Accreditation, 2009). The applicability of the biopsychosocial approach for the whole of psychology has also been explicitly recognized by the APA. In 2006, APA and 22 other US health care and human service professional organizations became signatories to the *Health Care for the Whole Person Statement of Vision and Principles*. This document is explicitly based on the biopsychosocial approach and emphasizes that health care and human services need to focus on the holistic functioning of individuals.

In addition to extensive support within psychology, the biopsychosocial approach has been widely integrated into health care and social services in general. The 23 signatories to the *Whole Person Statement* (APA, 2006) include organizations as diverse as the APA, American College of Obstetricians and Gynecologists, Society of Teachers of Family Medicine, American Nurses Association, American Public Health Association, and National Association of Social Workers. The biopsychosocial approach has been incorporated into the curriculum in nearly all medical schools in the United States and Europe (Frankel, Quill, & McDaniel, 2003) and is implied in the Accreditation Council for Graduate Medical Education (2011) requirements for medical residencies. It is also reflected in the standards of the Joint Commission on Accreditation of Healthcare Organizations (2006). This widespread adoption suggests that the biopsychosocial approach has become the standard of practice not only for behavioral health care but for health care in general in the United States.

The biopsychosocial approach for understanding human psychology is illustrated in Figure 7.1. The cube in that figure represents the three inextricably intertwined psychological, sociocultural, and biological dimensions of influence on human

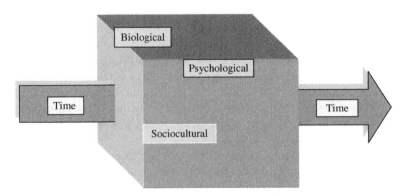

Figure 7.1 The biopsychosocial metatheoretical approach: human psychology is explained by the interactions between the three inextricably intertwined biopsychosocial dimensions across time.

psychology. This conceptualization also integrates the three levels of natural organization directly relevant to the individual human being, including the level just below (i.e., biological) and the level just above (i.e., the sociocultural). In addition, human development, functioning, and behavior change take place across time, and so it is necessary to incorporate a fourth dimension involving time, depicted in the figure by an arrow running through the biopsychosocial cube. This same conceptualization of human psychology applies to physical health and social behavior as well. Though physicians will usually focus on different aspects of the cube than will sociologists, this same framework applies across the medical, psychological, and sociological sciences. Understanding the complex nature of human development and functioning from any of these perspectives requires a comprehensive, systemic approach that spans all these levels.

The practice of psychology as conceptualized in this volume can also be depicted graphically. Figure 7.2 depicts an edifice resting on two pillars. The edifice represents the professional practice of psychology as a health care specialization, and this structure rests on two pillars representing scientific knowledge of human psychology and professional ethics. Together, science and ethics provide the pillars on which professional psychology is based. Remove either pillar and the foundations of professional practice become very unstable. In addition, understanding the practice of psychology as well as the underlying science and ethics all takes place within the context of the biopsychosocial perspective. The comprehensive understanding of all three of these factors requires an integrative biopsychosocial approach.

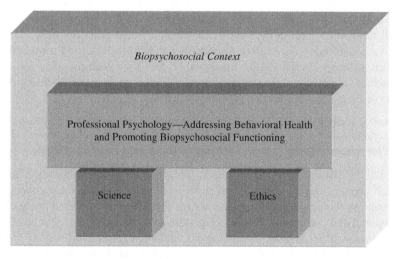

Figure 7.2 A unified conceptual framework for professional psychology: the professional practice of psychology rests on scientific knowledge of human psychology and on professional ethics; in addition, all these factors are understood within the context of the biopsychosocial approach.

Several of the specializations in professional psychology such as child, school, and health psychology, addictions, neuropsychology, and geropsychology already require a biopsychosocial approach to case conceptualization and treatment. As a result, clinical practices in those areas would be affected very little if the field as a whole adopted the biopsychosocial approach as its foundational conceptual framework.

Several other areas of professional psychology, however, and particularly the general practice areas, have not integrated this perspective as fully. This is probably the result of there being few specific proposals for how a biopsychosocial approach could be systematically applied in the general practice areas and across professional psychology as a whole. Medicine has considered this question at length (Frankel et al., 2003; White, 2005), but professional psychology has not. Therefore, the next section of this chapter discusses the primary implications of the above approach for professional psychology as a whole.

Implications of a Unified Biopsychosocial Framework for Professional Psychology

Some of the most important implications of the unified biopsychosocial framework described above for education and practice in the field are discussed below. To efficiently discuss these issues, it is useful to consider the context of graduate education in professional psychology. There is such great variety in health care systems across settings and specializations that discussing a biopsychosocial approach across those areas would be complicated. Students, faculty, and practitioners, however, all have some familiarity with standard educational practices in the field as a result of standard APA accreditation guidelines that are used across most programs. Therefore, to keep the discussion manageable, the basic features of a unified biopsychosocial framework for professional psychology are discussed below primarily in terms of their implications for graduate education.

1. *Instead of many different theoretical orientations for conceptualizing human development, functioning, and behavior change, there would be one metatheoretical framework for conceptualizing human psychology and the practice of psychology.*

From the perspective of the unified, science-based biopsychosocial approach described above, there is only one scientifically valid framework for conceptualizing human psychology and psychological practice. This framework emphasizes that the complexity of human psychology can be understood only through a comprehensive integration of biological, psychological, and sociocultural influences on development and functioning. This perspective applies at both the individual and population levels and from the perspective of both science and practice. So rather than ask professional psychology students to conceptualize cases according to a selected theoretical orientation, all students would learn to conceptualize cases according to a comprehensive, integrative biopsychosocial approach.

2. *Professional psychology curricula would be structured around learning the practice of psychology, not around learning particular theoretical orientations. From this perspective, the traditional systems of psychotherapy would generally be viewed as therapies and not as theoretical orientations.*

In addition to its scientific and ethical foundations, the practice of psychology includes core competencies in assessment, treatment planning, and the implementation of interventions to address the behavioral health needs of the general public. The curriculum for teaching these competencies would focus on a comprehensive biopsychosocial approach to assessment and treatment planning along with some range of interventions for treating the behavioral health problems faced by the general public. Using the unified science-based biopsychosocial approach, traditional systems of psychotherapy would be taught in the context of evidence-based treatments, but they would generally be viewed as *therapies* rather than as theoretical orientations. A science-based biopsychosocial approach would generally use the term *theory* and related terms in their scientific sense (which refers to explanations that have survived scientific tests aimed at verification and falsification), whereas *therapies* would be used to refer to treatments for addressing patient problems and concerns. The safety and effectiveness of therapies need to be established through experimental research, but therapies are not used as comprehensive scientific explanations of particular processes or phenomena.

Students in traditional professional psychology programs have often been expected to master one or a small number of theoretical orientations so that they can conceptualize cases and implement therapies consistent with that one (or more) orientation(s). This approach may be necessary for learning the individual therapies, but structuring training programs around learning theoretical orientations makes it difficult to specify the range of diagnostic and demographic populations with which students are competent to practice. For example, a student might learn to conduct cognitive therapy very capably with depressed middle-class Caucasian young adults who have no co-occurring conditions, but the student may not be competent to conduct cognitive therapy with depressed individuals across a range of demographic groups, those with co-occurring substance dependence or other Axis I disorders, Axis II disorders, or significant medical conditions, relationship dysfunction, parenting problems, or vocational instability. Working effectively with a range of depressed individuals requires the use of a comprehensive biopsychosocial approach that recognizes the importance of interacting sociocultural, biological, and psychological factors on individuals' functioning.

From the biopsychosocial perspective on professional psychology, establishing that students are competent to work with various clinical populations will include demonstrations of the ability to competently conduct assessment and treatment planning from a biopsychosocial orientation in addition to implementing some range of interventions that one can use to treat behavioral health issues. From this perspective, current language referring to theoretical orientations would generally be replaced by language referring to competencies and specializations. For example, rather than saying that one uses a cognitive behavioral theoretical orientation, one would note the populations one is competent to work with in terms of general or specialized practice and the particular therapies one is competent to provide. The number of different therapies that students can reasonably learn, as well as the types and range of therapies that are needed to meet the behavioral health needs of the public in different types of general and specialized practice, have not been studied extensively. These questions would receive more attention if a biopsychosocial approach to professional psychology education were adopted.

3. *Emphasis will be placed on conceptualizing patient cases in a comprehensive, holistic manner.*

The biopsychosocial approach involves a comprehensive, integrative perspective that considers the whole context of a person's development and functioning. As a result, a wide variety of psychologically, socioculturally, and biologically focused interventions can be incorporated within the biopsychosocial framework. Along with the traditional focus on psychological disorders, distress, and symptoms, a biopsychosocially oriented curriculum would also provide systematic coverage of a variety of issues related to effective functioning in important life roles in the family, at work, and in the community. For example, topics such as effective relationship and parenting skills, vocational effectiveness, general physical health, chronic health conditions, substance abuse, religion and spirituality, sexuality, and psychopharmacology have always been important in the practice of psychology, but they have often been learned largely outside the formal curriculum. A systematic biopsychosocial approach to professional psychology education would result in more attention to these topics within the curriculum. If students develop more comprehensive, integrative biopsychosocial case conceptualizations, they will also develop a stronger appreciation for the importance of collaborative interdisciplinary approaches to health care, another competency often considered critical to the practice of psychology (APA Assessment of Competency Benchmarks Work Group, 2007).

4. *The teaching of biological and sociocultural bases of behavior will be strengthened.*

The scientific understanding of biological and sociocultural influences on development and functioning has grown substantially in recent decades, and a science-based biopsychosocial orientation to behavioral health practice would emphasize these much more than have the traditional approaches to professional psychology education. For example, though multicultural competence is now widely recognized as critical for professional practice (APA, 2003), training programs still often include inconsistent coverage regarding the influence of childhood and family of origin environment, current family and other social support, education and employment, religion, class, and other important sociocultural factors on both normal and abnormal development. Possessing some significant and broadly based foundational knowledge in the biological bases of behavior (e.g., genetics, evolutionary psychology, the neurosciences, medical psychology) as well as the sociocultural bases of behavior is necessary for effectively implementing a biopsychosocial approach and for staying current with the rapidly emerging findings in these areas.

5. *General practice psychology will become better defined and delineated.*

When professional psychology education is focused more on the practice of psychology from a unified, science-based approach as opposed to the practice of particular therapies, more attention will be focused on the populations and disorders one is able to assess and treat. At present, the specializations (e.g., child, school, medical, and neuropsychology, psychopharmacology) tend to be organized around demographic and diagnostic groupings of patients and the competencies needed to meet their behavioral health needs. Training for the general practice of psychology (e.g., in clinical and counseling psychology), on the other hand, still often allows a great deal of latitude in terms of choosing one's theoretical orientation and the diagnostic and demographic groups with which one works.

"Broad and general preparation for practice at the entry level" is currently required as part of the APA standards for accredited professional psychology training programs (APA Commission on Accreditation, 2009, p. 3), but "broad and general preparation" is not further defined. A biopsychosocial perspective on general practice psychology focuses attention on meeting the behavioral health needs of the general public, which helps clarify the range of assessment, treatment planning, intervention, and other skills that program graduates and licensure applicants should possess in order to meet those needs. It also clarifies the relationship between general and specialized practice. At present, even very basic questions about whether general practice psychology should include working with children, adolescents, and seniors are not widely discussed (the assumption appears to be that general practice typically includes working with young and middle-aged adult outpatients). Greater focus on the practice of psychology, as opposed to the practice of particular therapies, quickly brings attention to these types of questions.

6. *Conflicts regarding the relevance of research to practice, the superiority of particular research methodologies or theoretical orientations, and other controversies that have fractionated the field will increasingly be viewed as outdated.*

The divides that have existed between clinicians and researchers, quantitative and qualitative researchers, and various theoretical schools and camps will gradually be replaced by common ground provided by a science-based biopsychosocial perspective. From this perspective, all levels of natural organization are critical to a complete understanding of human psychology, and all explanations are welcomed that shed light on the inextricably intertwined biological, psychological, and sociocultural influences on development, functioning, and behavior change.

Most of the contention in the field regarding the validity of various theoretical orientations and research methodologies is associated with the pre-paradigmatic era of psychology. In retrospect, theoretical orientations that focused on only particular aspects of the psychological, biological, and sociocultural domains would inevitably come into conflict because they necessarily provided incomplete explanations of human development and functioning. The science-based biopsychosocial approach advocated here, on the other hand, is deeply respectful of the complexity of human psychology, and consequently also skeptical of incomplete or reductionistic explanations of psychological phenomena. It is also automatically and continually updated because of the necessity to incorporate verified scientific findings into the framework. As a result, past conflicts about the validity of competing theoretical orientations will fade in importance.

The biopsychosocial approach and the scientific method also share similarities at practical and conceptual levels that will help researchers and clinicians avoid conflict and come together on a shared perspective of the discipline. For example, from a scientific perspective, one selects research methodologies for investigating particular phenomena based on the likelihood they will provide useful data and explanations, not on the basis of personal preferences or allegiances regarding which methodologies are the "best." Likewise, from a biopsychosocial practice perspective, one selects interventions to help patients with their problems and maximize their functioning based on the likelihood they will be effective in individual cases, not on the basis of personal preferences or allegiances regarding which interventions are the "best." The compatibility of these perspectives and the explicit recognition of the tremendous complexity of human psychology that is inherent in the biopsychosocial approach will shift attention away from the controversies of the past and toward advancing the scientific understanding of human psychology and its application in psychological practice.

Discussion

The field of professional psychology is ready to become paradigmatic. In fact, one might view the recent history of the field as teetering on the tipping point as scientific explanations for psychological phenomena grew in detail and comprehensiveness and some traditional practices were called into question. The heated controversy over the validity of recovered memories of child abuse in the 1990s (the "memory wars"; Loftus & Davis, 2006) might represent the period when the tipping point was reached. During that time, large numbers of therapy patients believed they had recovered memories of sexual and other forms of child abuse that they suffered as children. Heated debates ensued between practitioners who believed that the memories were accurate and researchers who argued that the reliability of remote childhood memories will often remain unknown in the absence of corroborating evidence. The empirical, scientific camp prevailed in that controversy and in others that followed (e.g., the scientific basis for rebirthing attachment therapy, the role of eye movements in eye movement desensitization and reprocessing therapy; see Chaffin et al., 2006; Perkins & Rouanzoin, 2002). These were followed by the formal endorsement of evidence-based practice principles that became APA policy in 2005.

This volume argues that there now appears to be consensus in the field that a biopsychosocial metatheoretical perspective can provide a common paradigmatic framework for professional psychology. There are many signs of agreement not just in psychology, but across the sciences, regarding the validity of this framework. Scientific explanations for specific psychological processes are also growing in number and detail, including for processes at medium and higher levels of complexity. It was argued in Chapter 5 that the field has reached the point where there are sufficiently detailed scientific explanations for enough psychological processes to justify a general transition to a unified science-based framework.

It should be noted that the choice of the biopsychosocial term is arbitrary in the sense that other terms (e.g., ecological, ecosystemic) also focus on the interactions between the relevant levels of natural organization in explanations of human psychology. The term *biopsychosocial*, however, is widely known throughout psychology, health care, and human service fields throughout the United States and internationally. It also incorporates the general levels just above and below the human individual, and is based on general systems theory, which was one of the important theoretical developments in complexity theory. Therefore, it is the best choice for representing a comprehensive science-based framework for understanding human psychology. It should also be noted that the arrangement of the biological, psychological, and social factors in the biopsychosocial term is essentially arbitrary. The ordering of these factors does not imply that biology is the most important level while the social domain is the least important (or vice versa). Instead, the ordering of the three domains represents increasing order of natural complexity.

If professional psychology does enter a new paradigmatic era, leaving behind the traditional theoretical orientations as central organizing frameworks for clinical practice will be controversial. Theoretical orientations have played a central role in professional psychology throughout its history, and replacing them would involve a major transition for many psychologists. This problem is mitigated somewhat by recalling the appropriate levels for conceptualizing different aspects of psychology. The biopsychosocial perspective approaches human psychology and behavioral health care at the overarching metatheoretical scientific level, whereas the traditional theoretical orientations have often been used to conceptualize specific psychological phenomena such as particular cognitive, affective, behavioral, developmental, pathological, and/or therapeutic processes. These orientations are inadequate as comprehensive scientific explanations for human psychology, though some orientations are supported by scientific explanations of particular psychological processes (e.g., behavioral therapy supported by the research on classical and operant conditioning, psychodynamic approaches supported by infant attachment research). Empirically supported treatments associated with the traditional theoretical orientations are easily integrated into a biopsychosocial framework, and therapists certainly can continue to use empirically supported interventions. Using these theoretical orientations as comprehensive explanations for human development and functioning is not supported by scientific evidence, however.

Keeping current with scientific advances and leaving behind the pre-paradigmatic era in professional psychology is critical for many reasons. Conflicts associated with competition between scientists and practitioners and between theoretical schools and camps have consumed large amounts of time and energy. At times, the conflict has been so strong that one could forget that much of the success of professional psychology is directly dependent on its scientific foundations. There is little doubt that the status of professional psychology would seriously erode if the scientific foundations of the field did not continue to strengthen. The ability of professional psychology to continue to prosper into the future is directly dependent on the strength of its scientific foundations.

Leaving behind the pre-paradigmatic era in professional psychology will have other benefits as well. One is simply that it serves the interests of the field to do so. Our credibility as a discipline within universities, our ability to attract the best students and obtain research grants, and our esteem as a profession within health care systems, institutions, and the government, and among the public generally are all affected by the strength of the scientific foundations of professional psychology and our effectiveness in applying that knowledge in professional practice. The conceptual confusion that characterized some of our past practices detracts from these interests. Medicine seems to have profited from leaving behind what could be considered a pre-paradigmatic approach to medical education and embarking on a more solidly science-based path following the adoption of the Flexner Report in 1910 (Sharpe & Faden, 1998). Professional psychology will also benefit from intentionally leaving behind our pre-paradigmatic past and transitioning to a unified science-based approach. Embracing a unified science-based conceptual framework for the practice of psychology will allow the field to move ahead with a unified voice for addressing the many psychological, social, and physical health problems that inhibit human welfare and potential.

Part III

Conceptualizing Psychological Treatment from a Biopsychosocial Perspective

Psychological intervention is undertaken to meet a wide variety of purposes. Some purposes are more focused and specialized, as in sports psychology, executive coaching, or various forensic contexts. Others are more comprehensive and general, as in many outpatient clinics. A biopsychosocial approach to professional psychology has distinct implications for conceptualizing the intervention process across all types of psychological practice. The chapters in this part of the book describe the basic implications of taking this approach across the four general phases of the treatment process, from assessment through treatment planning, treatment, and outcomes assessment. Many of these issues apply in nonclinical contexts as well, but the discussion here focuses on the behavioral health care treatment process.

There are no clear demarcations between the four phases of the treatment process reviewed in this part of the book. While the process begins with assessment, assessment continues throughout treatment—indeed, outcomes assessment is just the last phase of the ongoing assessment that occurs throughout treatment. Treatment also occurs throughout, from the first patient contact and the initial development of a therapeutic relationship through termination and the discussion of the outcomes of treatment. Nonetheless, it is necessary to divide the treatment process into its main phases because there are distinct shifts in purposes and activities as treatment proceeds.

It is important to note that the biopsychosocial approach described in this part of the book is consistent with the evidence-based practice approach that has been widely endorsed within health care in recent years (Institute of Medicine, 2001). The evidence-based approach to psychological practice integrates research evidence, clinical experience and judgment, and patient preferences and values (American Psychological Association Presidential Task Force on Evidence-Based Practice, 2006). The evidence-based practice and biopsychosocial approaches are quite compatible in that both are based on scientific research evidence but also rely on clinical experience for informing practice. The biopsychosocial approach acknowledges that scientific explanations of human psychology are growing in thoroughness but are still incomplete and consequently must be supplemented with clinical judgment. The role of the patient's sociocultural background, values, and preferences is certainly also prominent in the biopsychosocial approach to the treatment process.

8 Assessment

Psychological assessment is conducted for a wide variety of purposes. Some types of assessment serve nonclinical purposes, such as evaluations conducted for legal or administrative questions involving child custody, disability status, competency to stand trial, or the insanity defense. Assessments conducted for research purposes vary widely depending on the goals of the research. Within the clinical context, evaluations also vary greatly depending on whether they are conducted for emergency purposes (e.g., when patients are suicidal or homicidal), consulting purposes (e.g., to assist other treatment providers with their assessment and treatment planning), to reevaluate the progress and needs of patients in long-term care for the management of chronic conditions, or to conduct intake assessments to evaluate the needs of patients receiving behavioral health care for the first time.

The approach one takes to conducting psychological assessment depends not only on the specific purposes of the assessment but also on who requests the evaluation and the role of the psychologist in the future care of the patient. In some cases it is necessary to gather extensive information from those who requested the evaluation (e.g., when children are referred by parents or in many legal, academic, employment, or administrative evaluations). When the therapist expects to provide ongoing treatment to the patient, establishing an effective therapeutic relationship becomes a priority and may take precedence over the timely gathering of comprehensive assessment information.

Though the purposes and processes of psychological assessment vary greatly, several foundational issues underlie all types of assessment. This chapter focuses on these basic conceptual issues, with primary emphasis given to the behavioral health treatment context. To be sure, there are large bodies of specific knowledge and skill that must be acquired and applied to competently conduct psychological assessment, such as knowledge of normal development, personality, psychopathology, psychometrics, evidence-based practice, professional ethics and legal issues, as well as clinical interviewing and therapy relationship-building skills. This volume obviously does not address all these bodies of knowledge and skill. Instead, it focuses on the general conceptual framework that is used to conceptualize and understand behavioral health care from a biopsychosocial perspective.

The overarching purpose of professional psychology from the biopsychosocial perspective is to meet individuals' behavioral health needs while promoting their biopsychosocial functioning. Psychological assessment plays a critical role in this process, and this chapter examines the conceptual issues that underlie assessment from a biopsychosocial perspective. Specifically, it examines the basic reasons why assessment is conducted, the areas of patients' lives included in assessments, the

Foundations of Professional Psychology. DOI: 10.1016/B978-0-12-385079-9.00008-4

Table 8.1 Basic Issues in Conceptualizing Psychological Assessment from a
Biopsychosocial Perspective

- Overall purposes of psychological assessment
- Areas included in psychological assessment
- Sources of reliable and useful assessment information
- Thoroughness of the assessment information
- Assessing severity of patient needs
- Integrating assessment data
  - Prioritizing patient needs
  - Assessing overall complexity of patient needs
  - Integrating assessment information

sources for the most reliable and useful assessment information, the level of thoroughness of the assessment information that is needed, the evaluation of the severity of patient needs, and finally the integration of all the information collected (Table 8.1).

Overall Purposes of Psychological Assessment

The basic purposes underlying psychologists' approach to intake assessment and psychological evaluation have varied greatly over the years and across practitioners (for a historical overview, see Maloney & Ward, 1976; Tallent, 1992). Many practitioners have pursued relatively specific purposes associated with their personal theoretical orientation and have emphasized, for example, the family-of-origin experiences that underlie maladaptive personality characteristics, the importance of irrational thinking, the level of motivation for making behavior changes, or a family history of psychiatric illness that suggests neurochemical imbalances underlying psychological symptoms (e.g., Carr & McNulty, 2006; Eells, 2007; Wiggins, 2003). Others have pursued more general purposes aimed at gaining a comprehensive understanding of patients not tied to any of the traditional theoretical orientations.

To clarify the basic purposes of assessment in contemporary behavioral health care practice, several of the classic, influential, and official guidelines for conducting psychological assessment in the field are briefly noted below. The following guidelines capture a broad range of perspectives regarding the basic purposes for conducting psychological assessment:

- In their classic text *Psychological Assessment*, Maloney and Ward (1976) note that *"psychological assessment is a process of solving problems (answering questions)* in which psychological tests are often used as *one* of the methods of collecting relevant data" (p. 5; italics are in the original). They further explain that "To facilitate matters, we have broken down the process of psychological assessment into three discrete steps: problem clarification, data collection, and problem solution" (p. 8).

- In the *Handbook of Psychological Assessment* (5th ed., 2009), Groth-Marnat states that "The central role of clinicians conducting assessments should be to answer specific questions and aid in making relevant decisions. To fulfill this role, clinicians must integrate a wide range of data and bring into focus diverse areas of knowledge. Thus, they are not merely administering and scoring tests" (p. 3).
- With regard to neuropsychological assessment, Lezak (1995) notes that "Neuropsychological examinations may be conducted for any of a number of purposes: to aid in diagnosis; to help with management, care, and planning; to evaluate the effectiveness of a treatment technique, to provide information for a legal matter; or for research. In many cases, an examination may be undertaken for more than one purpose" (p. 110).
- In the *Use of Psychological Testing for Treatment Planning and Outcomes Assessment* (3rd ed., 2004a), Maruish notes that "Generally, psychological assessment can assist the clinician in three important clinical activities: clinical decision-making, treatment (when used as a specific therapeutic technique), and treatment outcomes evaluation" (p. 55). These three purposes are further specified as follows:
  - Clinical decision making includes three more specific functions:
    - Screening—"the use of brief instruments designed to identify ... the presence (or absence) of a particular condition or characteristic" (p. 55) that needs clinical attention.
    - Treatment planning—"Through their ability to identify and clarify problems as well as other important treatment-relevant patient characteristics, psychological assessments also can be of great assistance in planning treatment" (p. 55).
    - Treatment monitoring—"treatment monitoring, or the periodic evaluation of the patient's progress during the course of treatment" (p. 55).
  - Treatment—"In essence, assessment data can serve as a catalyst for the therapeutic encounter via (a) the objective feedback that is provided to the patient, (b) the patient self-assessment that is stimulated, and (c) the opportunity for patient and therapist to arrive at mutually agreed upon therapeutic goals" (p. 18).
  - Outcomes evaluation—"Psychological assessment can be employed as the primary mechanism by which the outcomes or results of treatment can be measured" (p. 56).
- Standards for psychological testing, ordinarily considered a subset of psychological assessment, emphasize similar purposes. For example, the *APA Guidelines for Test User Qualifications* (Turner, DeMers, Fox, & Reed, 2001) state that "Regardless of the setting, psychological tests are typically used for the following purposes:
  - *Classification*—to analyze or describe test results or conclusions in relation to a specific taxonomic system and other relevant variables to arrive a at classification or diagnosis.
  - *Description*—to analyze or interpret test results to understand the strengths and weaknesses of an individual or group. This information is integrated with theoretical models and empirical data to improve inferences.
  - *Prediction*—to relate or interpret test results with regard to outcome data to predict future behavior of the individual or group of individuals.
  - *Intervention planning*—to use test results to determine the appropriateness of different interventions and their relative efficacy within the target population.
  - *Tracking*—to use test results to monitor psychological characteristics over time" (p. 1104).

- According to *American Psychiatric Association Practice Guidelines for the Treatment of Psychiatric Disorders* (2006), The aims of a general psychiatric evaluation are:
  1) to establish whether a mental disorder or other condition requiring the attention of a psychiatrist is present;
  2) to collect data sufficient to support differential diagnosis and a comprehensive clinical formulation;
  3) to collaborate with the patient to develop an initial treatment plan that will foster treatment adherence, with particular consideration of any immediate interventions that may be needed to address the safety of the patient and others—or, if the evaluation is a reassessment of a patient in long-term treatment, to revise the plan of treatment in accordance with new perspective gained from the evaluation; and
  4) to identify longer-term issues (e.g., premorbid personality) that need to be considered in follow-up care (p. 7).

These guidelines clearly indicate the central role that assessment plays in behavioral health treatment. This initial phase of treatment is the primary point at which problems and concerns are identified and diagnosed. This greatly affects how the problems or concerns are understood by the patient, the therapist, and other stakeholders, and the type of services that are then provided to patients. This of course has a major impact on the subsequent course and outcome of treatment. If assessment is conducted improperly, problems can be missed or misidentified, the consequences of which can be serious. For example, a child's failure to succeed academically in school might be misattributed to a lack of motivation and effort rather than to a learning disability, discrimination the child is facing due to his or her sex, race, culture, or sexual orientation, or abuse or neglect the child is experiencing at home. Psychologists' ethical obligations to not cause harm, prevent foreseeable harms, and provide benefit can be violated if problems such as these are misdiagnosed.

The above guidelines indicate that psychological assessment serves multiple purposes beyond the initial identification of problems and concerns. While their emphases differ, there is significant convergence around what those purposes are. A synthesis of these guidelines and standards suggests that the basic purposes of psychological assessment are to:

1. Identify behavioral health problems and concerns that need clinical attention.
2. Gather information regarding a patient's behavioral health and biopsychosocial circumstances in order to develop a comprehensive case conceptualization and treatment plan.
3. Engage the patient in the treatment process through a collaborative approach that includes patient self-assessment and a discussion of objective feedback provided to the patient.
4. Provide ongoing assessment during treatment to monitor progress and refine the treatment plan and refocus interventions as needed.
5. Provide baseline data to conduct an outcomes evaluation to assess the effectiveness of treatment.

The specific purposes of assessment in particular cases obviously can vary substantially. Initial intake assessments with new patients are very different from the reevaluation for ongoing care of chronic issues with patients who are well known to the psychologist. Assessments conducted for consultation to others usually do not lead to one performing subsequent treatment or outcomes assessment at all. Across all types of psychological assessment, however, there is considerable convergence around the above general purposes of psychological assessment.

The sections below discuss the issues that need to be addressed in order to achieve the above basic purposes of psychological assessment. This examination starts with a consideration of which areas of patients' lives need to be assessed in order to develop comprehensive case conceptualizations from a biopsychosocial perspective.

Areas Included in Psychological Assessment

Consensus regarding the need to take a comprehensive biopsychosocial approach for conceptualizing psychological assessment appears to have been reached in recent years. A survey of our standard textbooks and guidelines for learning assessment (such as those referenced in the previous section) indicate widespread agreement that psychologists should evaluate a full range of psychological, sociocultural, and physical health issues when conducting assessments. This was not always the case. In years past, therapists often approached assessment and case conceptualization based on the dictates of their particular theoretical orientation—the findings of assessments often followed directly from the theoretical approach used to conceptualize the case (e.g., a patient's depression could have been caused by an unresolved oral fixation, depressogenic cognitions, conditions of worth imposed by parents, or an enmeshed family system; Garb, 1998).

The Diagnostic and Statistical Manual of Mental Disorders (DSM) provides a clear example of how case conceptualization has evolved over the decades. The first edition of the DSM, published in 1952, relied heavily on psychodynamic theory with regard to many of the diagnostic categories. Symptoms for specific disorders were not specified in detail, and many were seen as reflections of broad underlying conflicts or reactions to life problems that could be categorized generally as either neurosis or psychosis. With the rise of alternative theoretical explanations for psychological development in the 1950s through the 1970s (e.g., humanistic, cognitive, feminist, biological, and sociological approaches), the weaknesses of the DSM-I and DSM-II became obvious.

A transformation in the conceptualization of psychiatric diagnosis occurred with the publication of the third edition of the DSM in 1980 (DSM-III; American Psychiatric Association, 1980a). This revision proved to be much more useful than earlier editions due to the use of an atheoretical descriptive approach that did not specify or imply etiology for most of the disorders. This revision also introduced the multiaxial assessment system, which, with modifications, is still in use today in the DSM-IV-TR (American Psychiatric Association, 2000a). The multiaxial assessment system incorporated what is essentially a biopsychosocial approach to assessment. The current five-axial system includes clinical disorders and conditions on

Axis I, personality disorders and pervasive developmental problems on Axis II, medical issues on Axis III, environmental stressors on Axis IV, along with general, overall level of functioning on Axis V.

The DSM-III five-axial assessment approach represented a major improvement in broadening the assessment of mental health and biopsychosocial functioning. It provides little guidance, however, regarding the breadth and specificity of the factors that need to be included when attempting to understand development and functioning. Though the three general biopsychosocial domains are included in the system, the specific factors that need to be incorporated into diagnosis and assessment are not specified. Because it is a descriptive system with little emphasis on etiology, the five-axial diagnosis also provides little guidance on how to understand the causes of patients' problems or the significant risk and protective factors that affect the individual's current functioning. Much more information is needed to conduct comprehensive and useful assessments that can identify solutions to problems in individual cases (e.g., American Psychiatric Association, 2006; Beutler, Malik, Talebi, Fleming, & Moleiro, 2004; Goodheart & Carter, 2008).

To clarify the areas of patients' lives that should be considered in psychological assessment, several influential approaches to conducting comprehensive behavioral health assessment are briefly examined below (see also Meyer, 2008). Though there is variability across these systems, they also reflect substantial agreement regarding many of the components that should be incorporated into behavioral health assessments. Table 8.2 notes the specific components that are addressed by these six assessment systems. This first system is the *Addiction Severity Index* (McLellan et al., 1992), which is probably the most frequently used assessment instrument in the addictions field. It involves a semistructured clinical interview for obtaining systematic biopsychosocial data from patients. The American Psychiatric Association (2006) published a new edition of the *Practice Guidelines for the Treatment of Psychiatric Disorders: Compendium 2006* that addresses three types of evaluation: general psychiatric evaluation, emergency evaluation, and clinical consultation. For present purposes, the general psychiatric evaluation guidelines are examined here (the American Psychological Association has not yet published a similar set of practice guidelines). Campbell and Rohrbaugh (2006) published the *Biopsychosocial Formulation Manual* to document the model they use to train psychiatry residents. Groth-Marnat (2009) published the popular and well-respected *Handbook of Psychological Assessment*, now in its fifth edition. The next assessment framework considered is highly influential as a result of being promulgated by the primary accrediting body for hospitals and outpatient healthcare facilities in the United States. The Joint Commission on Accreditation of Healthcare Organizations (JCAHO, 2006) developed the *Provision of Care Standards* that must be met during the intake process by inpatient or outpatient behavioral health care facilities to receive JCAHO accreditation. Sperry (1988, 1999, 2001, 2006) has long advocated for a biopsychosocial approach to behavioral health treatment. His model, called biopsychosocial therapy, is an integrated approach that customizes treatment to the individual patient and is considered especially useful for complex and difficult cases.

Table 8.2 Components Addressed by Selected Mental Health Assessment Frameworks

Domains (in bold) and Components	Addiction Severity Index	American Psychiatric Association Practice Guidelines	Biopsychosocial Formulation Manual	Groth-Marnat Assessment	JCAHO Provision of Care Standards	Sperry Biopsychosocial Therapy
Biological						
General medical history	✓	✓	✓	✓	✓	✓
Childhood health history		✓	✓	✓		
Medications	✓	✓	✓		✓	
Health habits and behaviors	✓	✓			✓	
Psychological						
History of present problem	✓	✓	✓	✓	✓	
Individual psychological history	✓	✓	✓	✓	✓	✓
Substance use and abuse	✓	✓	✓	✓	✓	
Suicidal ideation and risk assessment	✓	✓	✓	✓	✓	
Individual developmental history		✓	✓	✓		✓
Childhood abuse history	✓	✓	✓			
Other psychological traumas		✓	✓	✓		✓
Mental status examination		✓	✓	✓	✓	✓
Personality style and characteristics			✓			
Sociocultural						
Relationships and support system	✓	✓	✓	✓	✓	✓
Current living situation	✓		✓	✓	✓	
Family history	✓	✓	✓	✓		✓
Educational history	✓	✓	✓	✓		
Employment	✓	✓	✓	✓		
Financial resources	✓	✓	✓	✓	✓	✓
Legal issues/crime	✓	✓	✓	✓		
Military history		✓			✓	
Activities of interest/hobbies	✓	✓	✓	✓	✓	
Religion[a]	✓		✓		✓	✓
Spirituality[a]			✓		✓	✓
Multicultural issues	✓	✓		✓		

[a] Religion here refers to organized religious practices and activities, whereas spirituality focuses on personal beliefs and meaning that may or may not include a "higher power" or organized religious practices.

Note: This table is adapted from Appendix A in Meyer (2008).

This table indicates clear agreement regarding the importance of all three of the general biopsychosocial domains to psychological assessment, though there is significant variability in the specific components included in each assessment system. Nearly all of the components listed in Table 8.2 are addressed by at least three of these systems. Taken together, the above listing suggests that all the components could be considered at least potentially important in psychological assessment.

These assessment systems consistently emphasize problems, however, with comparatively little emphasis on the assessment of strengths, resources, and assets. Rarely are clinicians directed to obtain information on strengths as well as deficits in each area. This is not unexpected, however. Until recently in human history, health care generally and mental health care specifically were focused on the assessment and treatment of problems for obvious reasons. Strengths, resources, and assets have always been important in people's lives, but the critical impact of disease, injury, and disability required that health care focus on treating problems as opposed to evaluating and developing strengths.

Over the past century, however, much of the world has entered a new era characterized by less serious disease and disability and far longer life spans. This represents a truly revolutionary shift in the life experiences for large proportions of the population in many countries around the world. In general, these involve a shift from focusing on meeting basic physical needs to meeting psychological ones. John Maynard Keynes (1930, p. 328) expressed the challenge as how "to live wisely and agreeably and well" once desperation and deprivation are no longer the driving forces of human existence. Humanistic psychologists also elaborated on the importance of these issues. Abraham Maslow (1943) conceptualized individuals' biopsychosocial needs in terms of a hierarchy where one's basic material and security needs had to be met before the needs for feeling loved, a sense of belonging, and self-esteem could be met. After these needs were met, one could focus on self-actualization and the pursuit of meaning and fulfillment in life. Carl Rogers (1961) described similar aspirations in terms of becoming a "fully functioning" person. More recently, Keyes (2007) quantified aspects of these concepts and found that only about 2 in 10 Americans are actually "flourishing," functioning at an optimal psychological level without significant distress, lack of meaning and purpose, or physical disability.

This evolution in the priority of basic physical versus psychological needs has major implications for psychological development and behavioral health and consequently must be considered in psychological assessment procedures as well. The biopsychosocial approach is based on a comprehensive, integrative approach to understanding development and functioning, and the full range of biopsychosocial functioning must be included in psychological assessment as a result. This requires a focus on strengths as well as weaknesses. For example, what people are doing well or possess in terms of a strength frequently has as much significance in their lives as what they are doing poorly or what they lack (e.g., as when a person who has had little success with intimate relationships also has rewarding friendships and is a conscientious and highly valued employee; or when an individual who has conflictual, disruptive relationships with his parents and siblings but has very satisfying and cooperative relationships at work and with his spouse and children). Focusing on only a person's maladaptive characteristics and behaviors can provide a very limited picture of a person's life and lead to seriously incomplete assessment results.

Taking these various perspectives into consideration, the areas of a person's life that should be evaluated in psychological assessment include problems, deficits, and disorders along with strengths, resources, and assets across the full range of biological, psychological, and sociocultural functioning. Building on the findings summarized in Table 8.2, Table 8.3 lists components to include in psychological assessment along with brief summaries of the content of each component. A recent study by Meyer and Melchert (2011) examined the content-related validity of this conceptualization of psychological assessment and found that the components listed in Table 8.3 were able to capture 100% of the intake information that was found in a sample of 163 individual therapy outpatient files from three different clinics. The content of each file was analyzed and categorized into these 26 areas (for purposes of clarity, the "Health Habits and Behaviors" component is separated out in Table 8.3, whereas it had been subsumed under "General Medical History" in the Meyer and Melchert study; and the "Level of Psychological Functioning" component is also separated out in Table 8.3, whereas it had been subsumed under "History of Present Problem" in the Meyer and Melchert study). There was no information found in any of the study files that could not be categorized into these component areas.

All of these areas of patients' lives clearly can be important in their development and current functioning. As a result, therapists need to assess each of these areas to gain a comprehensive understanding of patients' behavioral health and life circumstances. The depth and detail that one pursues in particular assessments depends on one's specialization, the setting where one practices, and the purpose of the assessment. For purposes of developing a general framework for conceptualizing psychological assessment, however, the above categorization provides a useful delineation of the areas of patients' lives encompassed in a biopsychosocial approach to assessment.

Sources of Assessment Information

After the primary purposes and focus of a psychological assessment have been identified, it is important to consider the best sources for obtaining the needed assessment information. Patients obviously present with a wide variety of problems and concerns. Often they are self-referred as a result of distress or concern they feel regarding a personal issue and the patient himself or herself provides most or all of the information needed for the assessment. At other times, parents, spouses, partners, employers, physicians, or educators initiate the referral and may provide information that is central to the assessment. The nature of the specific case determines which sources of information will provide the most reliable, relevant, and useful information in that case.

Table 8.3 Biopsychosocial Component Areas of Psychological Assessment

Domains (in bold) and Components	Issues Commonly Included
Biological	
General medical history	Current medical functioning, recent and past medical history, chronic medical illnesses, nondiagnosed health complaints, physical disability, previous hospitalizations, surgery history, seizure history, physical trauma history
Childhood health history	Birth history, childhood illness history, childhood psychiatric history
Medications	Dosage, efficacy, side effects, duration of treatment, medication adherence
Health habits and behaviors	Diet and nutrition, activity, exercise
Psychological	
Level of psychological functioning	Overall mood and affect, level of distress, impairment in functioning
History of present problem	Chronological account of recent symptoms, exacerbations or remissions of current illness or presenting problem, duration of current complaint, reason for seeking treatment at this time, previous attempts to solve the problem, treatment readiness (motivation to change, ability to cooperate with treatment)
Individual psychological history	Current psychiatric problems, previous diagnoses, treatment history (format, frequency, duration, response to treatment, satisfaction with treatment)
Substance use and abuse	Types of substances used (alcohol, tobacco, caffeine, prescribed, over-the-counter, illicit), quantity and frequency of use, previous treatments, other addictive behaviors
Suicidal ideation and risk assessment	Intent, plan, previous attempts, other self- and other-destructive behaviors (e.g., injury to self, neglect of self-care, homicidal risk, neglect of children or other dependents)
Individual developmental history	Infancy, early and middle childhood, adolescence, early and middle adulthood, late adulthood
Childhood abuse history	Physical, sexual, and emotional; psychological response to abuse
Other psychological traumas	Traumas and stressful life events, exposure to acts of war, political repression, criminal victimization
Mental status examination	Orientation, attention, memory, thought process, thought content, speech, perception, insight, judgment, appearance, affect, mood, motor activity
Personality style and characteristics	Coping abilities, defense mechanisms, problem-solving abilities, self-concept, interpersonal characteristics, intrapersonal characteristics

(Continued)

Table 8.3 (Continued)

Domains (in bold) and Components	Issues Commonly Included
Sociocultural	
Relationships and support system	Immediate and extended family members, friends, supervisors, coworkers or other students, previous treatment providers, current parent–child relationship, involvements in social groups and organizations, marital/relationship status and history, recurrent difficulties in relationships, presence of past and current supportive relationships, sexual and reproductive history
Current living situation	Current living arrangements, satisfaction with those arrangements
Family history	Family constellation, circumstances, and atmosphere; recent problems with family; family medical illnesses, psychiatric history and diagnoses; history of suicide in first- and second-degree relatives, family problems with alcohol or drugs, loss of parent and response to that loss
Educational history	Highest level completed, profession or trade skills
Employment	Current employment, vocational history, reasons for job changes
Financial resources	Finances and income
Legal issues/crime	Current legal issues and criminal victimization, legal history
Military history	Positions, periods of service, termination
Activities of interest/ hobbies	Leisure interests and activities, hobbies
Religion	Organized religious practices and activities, active in faith
Spirituality	Personal beliefs and meaning (which may or may not include a "higher power" or organized religious practices)
Multicultural issues	Race/ethnicity, racial/ethnic heritage, country of origin

Source: Table is adapted from Appendix E in Meyer (2008).

The reliability of the information used to inform psychological assessment is critical in clinical practice. Reliable assessment information can be efficiently collected through patient verbal self-report with regard to some topics, while other issues are often most reliably and efficiently assessed through the use of questionnaires, screening instruments, or psychological tests. For example, one's level of distress, mood, and other subjective states often can be assessed only through self-report. Various psychological variables such as personality characteristics, educational achievement, and intellectual or neuropsychological functioning often are most reliably and efficiently assessed through the use of test instruments. When it comes to evaluating a patient's performance of responsibilities at home or at work, on the other hand, family members and work supervisors often provide more reliable and complete information than what patients themselves might be aware of or willing to report. Children and cognitively disabled adults are usually unable to provide reliable reports regarding several aspects of their lives. Legal, medical, substance abuse, educational, and child protective service issues also may not be reliably reported by patients themselves. Though patient self-report information is often the most time-efficient to collect, it often carries an unacceptably high risk of being seriously incomplete or inaccurate.

In general, patients themselves are the primary source for information about their personal distress and other internal states they experience. Therapists often have the most expertise to reliably identify psychological symptoms and make psychiatric diagnoses. A patient's medical status ordinarily is best understood by his or her medical providers, while family members often have the most insight regarding the patient's functioning within the family. Employers or educators often have the best perspective on a patient's functioning at work or school, while officials within criminal and legal systems often can provide reliable information regarding patients' legal involvement.

A useful model for organizing sources of assessment information was proposed by Strupp and Hadley (1977). Their *Tripartite Model of Mental Health and Therapeutic Outcomes* noted that at least three different stakeholders hold different perspectives and have different interests in a patient's psychological functioning and treatment. First, they argued that the patient is the best judge of his or her own distress and discomfort. Second, the patient's family and community have the best perspective for judging a patient's functioning in his or her important life roles (e.g., within the family, at work, in the community). Third, therapists are normally the best judges of changes in patient's psychological functioning and psychopathology. Strupp and Hadley argued that the comprehensive assessment of the effectiveness of treatment should include all three of these perspectives.

Speer (1998) expanded on the Strupp and Hadley (1977) model by specifying the sources of information that are likely to provide the most useful and reliable information for the comprehensive assessment of a patient's functioning, adjustment, and treatment outcomes (Table 8.4). In this model, significant others include employers, neighbors, friends, and landlords in addition to family members. Public gatekeepers are those who have professional responsibilities involving the patient but not a social relationship with him or her, such as law enforcement officials, emergency room staff, court officials, and child or adult protective services staff. Independent observers are professionals or specialists who can perform medical, psychiatric, or other evaluations of the patient. The capitalized bold letters in Table 8.4 indicate those individuals who are likely to provide more reliable information with regard to the different dimensions of psychological adjustment and functioning.

The importance of obtaining reliable assessment information is evident when one considers the frequency with which different informants provide completely different perspectives on the issues that patients present. For example, a husband entering treatment might ask for help with getting along with a "nagging" spouse, while the spouse might report that the husband's alcohol abuse is about to result in a divorce and child custody battle. A patient might report that an angry, demanding

Table 8.4 Reliable Sources of Assessment Information (from Speer, 1998, p. 50)

Source	Distress	Symptoms, Disorder, Diagnosis	Functioning, Role Performance
Patient	A	B	c
Significant others	d	e	F
Public gatekeepers	g	h	I
Independent observers	j	K	l
Therapist/provider	m	N	o

Note: Bold capital letters indicate sources more likely to provide reliable information.

supervisor at work is unreasonable and unfair, but the supervisor might report that the employee has engaged in sexually harassing behaviors, has frequent conflicts with coworkers, and has substandard productivity in both quantity and quality. Relying only on patient self-report in these cases obviously can result in seriously inaccurate assessments that might not only be unhelpful but could also result in negative consequences for the patient or others.

Unreliable and incomplete assessments can lead to interventions that are unhelpful or even deleterious. Therefore, professional psychologists need to be adept at collecting data from a variety of sources to obtain the most relevant and reliable assessment information possible. This also requires the ability to work collaboratively with other human service professionals, family members, and significant others.

Thoroughness of the Assessment Information

The previous two sections considered the breadth of information to be collected in psychological assessments along with sources that can provide reliable and useful information. This information also needs to be sufficiently thorough and complete before it can be evaluated and incorporated into a psychological assessment. Information that is missing or incomplete regarding important factors in patients' lives obviously can result in inaccurate assessment findings and treatment recommendations. As mentioned in Chapter 6, the patient safety movement in American medicine over the past decade has raised concern about the impact of missed and delayed diagnosis (Wachter, 2009). The Institute of Medicine in 2000 estimated that 44,000–98,000 Americans die each year as a result of medical errors ("a jumbo jet a day"). While many of these errors are associated with factors such as medications and infections acquired while receiving health care, missed and delayed diagnoses also account for many deaths. Most of these diagnostic errors involve nonpsychiatric issues, but depression with subsequent suicide attempt may be among the common missed diagnoses (Schiff et al., 2009).

There is an overwhelming amount of patient information that could be collected across the biopsychosocial domains. Fortunately, clinical experience leads to therapists becoming increasingly efficient by focusing on the most relevant and salient information for the purposes involved. In general, assessments tend to be more comprehensive and detailed when patients' concerns and problems are more serious and complex. The setting and specialization in which one practices also has a major impact on the thoroughness of the information collected. Inpatient programs routinely require that medical and psychosocial evaluations be completed to address the severity and complexity of the issues involved. On the other hand, employee assistance programs, university counseling centers, and school counseling departments generally use brief screening approaches as a result of the number of employees or students covered and the level of treatment that can be provided.

The use of standard intake questionnaires and interview protocol forms can help ensure that the collection of assessment information is thorough. The use of standardized screening instruments has also become widely recommended recently because of their usefulness for providing psychometrically reliable and valid data along with normative data. These instruments can also be readministered during and after treatment, thereby providing a very useful mechanism for monitoring treatment progress and evaluating outcome (see Chapter 11).

At a very basic level, the adequacy and thoroughness of the assessment information collected for a given patient case can vary from completely inadequate (e.g., almost nothing is known about important relevant issues) to fully adequate for the purpose. Because of variation in the purposes of assessment and the uniqueness of each patient case, it is not possible to establish precise guidelines regarding what would qualify as adequate and thorough assessment. To identify the level of thoroughness of assessment data at the general outpatient level, however, Meyer and Melchert (2011) developed a five-point scale to rate the thoroughness of assessment data for each of the individual biopsychosocial component areas listed in Table 8.2. The descriptors for each of the five points on the rating scale are noted in Table 8.5. Given the usefulness and importance of assessing strengths in addition to deficits for gaining a thorough understanding of a patient's circumstances, these are both incorporated into the scale as well. To illustrate the application of this approach, Table 8.6 provides examples of intake assessment notes for each of the five levels on the scale. More thorough assessment information is clearly important for conceptualizing cases in a more detailed and comprehensive manner that subsequently also increases the likelihood of treatment effectiveness.

Assessing Severity of Patient Needs

In addition to collecting thorough information regarding patients' problems, it is critical to evaluate their severity because of the direct implications of problem severity for treatment planning. For example, serious emergency needs must be attended to immediately, regardless of whether they involve psychological

Table 8.5 Detail and Comprehensiveness Scale for Assessing Biopsychosocial Components

Score	Rating Description
0	Information regarding component area is not present at all.
1	Only a few details or basic data are mentioned; or a checkbox for this component is marked but no further information is provided.
2	Most or nearly all basic details or data are present; strengths and/or weaknesses may be mentioned minimally, but not clearly assessed as a strength or a deficit.
3	Most or nearly all details or data are present plus one of the following two are also met: 1) strengths associated with this component are described, or 2) deficits associated with this component are described.
4	All of the following criteria are met: 1) most or nearly all details or data are present; 2) strengths associated with this component are described; and 3) deficits associated with this component are described.

Source: Adapted from Appendix F in Meyer (2008).

(e.g., suicidality), medical, family, legal, or other issues. Other needs may be quite serious but not urgent, and will require intensive intervention but not on an emergency basis. On the other hand, other needs are minor and can be addressed through psychoeducation or a referral to external sources of information or support. Patients also often enjoy high levels of functioning or strong support in particular areas of their lives, which can serve as important sources of stability when addressing problems and needs in other areas.

Many models for assessing the severity of patients' needs range from "none" to "severe." The DSM system has used "mild," "moderate," and "severe" to indicate level of severity of mental disorders since the third edition (American Psychiatric Association, 1980), and many other systems have incorporated these same terms and concepts (e.g., Huyse et al., 2001).

In addition to noting the severity of problems and disorders, a biopsychosocial perspective on assessment emphasizes the importance of positive functioning and personal resources as well. As noted earlier, behavioral and medical health assessment in the past tended to emphasize deficits and pathology to address patient needs. In addition, many therapists also operated from the perspective of a particular theoretical orientation, and there may have been a tendency to notice or look for problems that could be treated from that orientation. A biopsychosocial perspective to health care, on the other hand, emphasizes the whole person and the full continuum of functioning across areas of development and functioning. Conducting a holistic biopsychosocial assessment requires a comprehensive assessment of strengths and resources as well as problems and needs. Strengths and resources include both internal resources (e.g., coping skills) and external resources (e.g., social support).

Table 8.7 illustrates how the full continuum of need severity can be conceptualized. Rather than conceptualizing problems using a unipolar scale that ranges from no problem to severe problem (e.g., the DSM approach), a bipolar scale can also

Table 8.6 Examples of Intake Assessment Notes Documenting Particular Assessment Issues

Score	Substance Use Example	Medication Example	Religion Example
0	[Information regarding this component area is missing]	[Information regarding this component area is missing]	[Information regarding this component area is missing]
1	"Patient states she drinks alcohol."	"Patient takes Prozac."	"Patient is Roman Catholic."
2	"Patient reports drinking alcohol socially, approximately twice per month. She reports not smoking and does not consume caffeine or any illicit drugs."	"Patient currently takes Prozac, 40 mg, once daily for Depression."	"Patient is Roman Catholic, is active in her faith, attends Church regularly, and was raised as a Catholic."
3	"Patient reports drinking alcohol socially, approximately twice per month. She reports not smoking and does not consume caffeine or any illicit drugs. Patient reports drinking has negative impact because when she goes out and drinks with friends, she usually drinks too much and does not want to get out of bed the next day."	"Patient currently takes Prozac, 40 mg, once daily for Depression; he states that the medication is helpful because he no longer feels depressed and is more active socially."	"The patient reports that she is Roman Catholic, is active in her faith, goes to church regularly, and was raised Catholic; she states that her religion has helped her by providing a positive support group during her recent difficulties."
4	"Patient reports drinking alcohol socially, approximately twice per month. She reports not smoking and does not consume caffeine or any illicit drugs. Drinking on a social basis has been helpful, according to the patient, because she gets to go out with friends and feels more comfortable socializing and meeting new people. Patient reports that drinking has a negative side effect as well because when she goes out and drinks with friends, she usually drinks too much and does not want to get out of bed the next day."	"Patient currently takes Prozac, 40 mg, once daily for Depression; he states that the medication is helpful because he no longer feels depressed and is more active socially; he reports the medication has a downside as well—he is afraid that he will have to take the medication 'forever.'"	"The patient reports that she is Roman Catholic, is active in her faith, goes to church regularly, and was raised Catholic; she states that her religion has helped her by providing a positive support group during her recent difficulties. Her religion has had a detrimental effect as well, though, because she states that she does not always agree with Church doctrine, and feels a great deal of internal conflict and guilt as a result."

Source: Adapted from tables 3.3, 3.4, and 3.5 in Meyer (2008).

Table 8.7 Assessing Severity of Need and Strength of Resources Across Biopsychosocial Areas

−3	**Severe need**—patient is functioning far below an optimal level and/or risks a major deterioration in level of functioning with dangerous or disabling consequences possible
−2	**Moderate need**—patient is functioning significantly less than optimally and/or is facing risks for a significant deterioration in level of functioning
−1	**Mild need**—patient is experiencing mild psychological distress and/or impairment in functioning or faces minor risks for a decline in functioning
0	**No need**—no evidence of need in this area, though also not an area of strength
+1	**Mild strength**—a mild strength or resource for the patient; may be developed and amplified further
+2	**Moderate strength**—a moderate strength or resource that adds significantly to the patient's health and functioning; may be developed or amplified further
+3	**Major strength**—a major strength or resource that is an important contributor to the health and well-being of the patient

incorporate the positive dimensions of functioning in particular areas of an individual's life. Therefore, the scale depicted in Table 8.7 ranges from severe need at the low end to major strength at the high end. This conceptualization does not apply neatly to all areas of biopsychosocial assessment. For example, if one had no significant childhood illnesses or injuries, it is unclear whether that is best viewed as a strength or simply as having no needs in that area. If a person with a history of child abuse or neglect has worked through the consequences of those experiences and has developed strong resiliency and valuing of healthy relationships as a result, these consequences of the experience of child abuse could be viewed as a strength. If used as a measurement model, all these issues would need careful analysis and evaluation. As a conceptual model, however (which is our interest here), a bipolar conceptualization of problem severity is very useful as a reminder to assess both strengths and needs across biopsychosocial areas.

Table 8.8 illustrates how a bipolar biopsychosocial conceptualization of patient needs is useful for gaining a thorough assessment of a patient case that can then lead to clear and well-supported treatment plans. For example, the dots in the table summarize the assessment of needs and strengths across the biopsychosocial areas for the case of a depressed business executive who is very effective at work, managing a large number of important responsibilities with positive appraisals by the chief executive. This patient has distant and perfunctory relationships with his wife and children, however, as well as distant and conflictual relationships with his parents. He consumes significant amounts of alcohol when he is not working, apparently to help avoid the emptiness and anger he feels regarding his personal and social life. He also neglects his physical health. As another example, the checks in the table refer to the global assessment of the needs and strengths of a homeless man with bipolar affective disorder and substance dependence. He has a pleasant and engaging personality but has significant needs and problems in most areas of his life.

Table 8.8 Example of Biopsychosocial Assessment for Two Cases: A Business Executive and Homeless Individual (Dots and Checks, Respectively)

Biopsychosocial Domains and Components	−3 Severe need	−2 Moderate need	−1 Mild need	0 No need	+1 Mild strength	+2 Moderate strength	+3 Major strength
Biological							
General physical health		✓	●				
Childhood health history		✓	●				
Medications		✓	●				
Health habits and behaviors		✓	●				
Psychological							
Level of psychological functioning		✓●					
History of present problem	✓	●					
Individual psychological history	✓		●				
Substance use and abuse	✓	●					
Suicidal ideation and risk assessment		✓	●				
Effects of developmental history		✓	●				
Childhood abuse and neglect	✓		●				
Other psychological traumas		✓		●			
Mental status examination			✓				●
Personality style and characteristics			●			✓	
Sociocultural							
Relationships and social support	✓	●					
Current living situation	✓					●	
Family history	✓	●					
Educational history		✓					●
Employment	✓						●

(Continued)

Table 8.8 (Continued)

Biopsychosocial Domains and Components	−3 Severe need	−2 Moderate need	−1 Mild need	0 No need	+1 Mild strength	+2 Moderate strength	+3 Major strength
Financial resources	✓						●
Legal issues/crime			✓		●		
Military history			✓	●			
Activities of interest/ hobbies		✓●					
Religion					✓●		
Spirituality			✓●				
Multicultural issues				✓●			

Therapists do not always conduct detailed assessments covering all of these areas because brief screening is sufficient for several purposes. In general, however, it is important to include a bipolar conceptualization of functioning rather than focusing only on problems and pathology because of the importance of gaining a thorough understanding of a person's circumstances and functioning. It is difficult to develop individualized treatment plans that address each patient's particular circumstances without such a conceptualization. This type of individualized approach that includes strengths as well as problems also communicates to patients that the therapist is interested in them as whole persons, and not just interested in their problems (or interested in them only when they have problems). This in turn helps develop rapport and a stronger therapeutic relationship, which is also important to positive treatment outcomes (see Chapter 9).

Integrating Assessment Data

Use of the above guidelines will lead to thorough information being collected for comprehensive biopsychosocial assessment. A critical step remains, however, because the information collected needs to be integrated and organized in a holistic manner designed to maximize the likelihood of treatment effectiveness. To achieve this type of assessment, three additional issues need to be evaluated: (1) the prioritization of the patient's needs; (2) the complexity of the patient's needs taken together as a whole; and (3) the clinically useful integration of all the assessment information gathered.

Prioritizing Needs

The identification of needs that require immediate, intensive intervention is often not complicated. By definition, emergency needs fall into this category. In mental health care, the most common emergencies involve danger to self or others. Sometimes therapists encounter other types of emergencies, such as medical, family, legal, criminal, or other issues, and resolving those needs is often the first priority before other needs are addressed. For example, attempting to resolve a suicidal college student's career indecision before the suicidality has been adequately resolved may not only be unhelpful but has the potential to increase stress and uncertainty and consequently also the chances of a suicide attempt. Therefore, prioritizing patient needs is essential for intervention to proceed in a therapeutic manner.

The best known approach to conceptualizing the prioritization of human needs is Maslow's (1943) Hierarchy of Needs model (Figure 8.1). Maslow considered the four lowest levels of needs (physiological, safety, love/belonging, esteem) to be "deficiency" needs—only when they are met can the individual move up the hierarchy and establish new priorities for personal growth. Prioritizing needs is critical to integrative assessment and the development of appropriate treatment plans. For example, homeless individuals worried about basic needs for clothing, food, shelter, or physical safety may find it impossible to focus on higher-level needs until some level of basic physical stability is achieved. Focusing on existential issues regarding meaning and fulfillment in life can be very difficult and perhaps even counterproductive if one's basic needs for social connection and self-esteem have not been met.

Research has shown that need fulfillment is more fluid than that suggested by a stepped hierarchical model (Wahba & Bridgewell, 1976). Nonetheless, Maslow's model is widely considered useful for categorizing different types of needs and arranging their priority. Especially when patients have limited insight into the nature and interrelationship of their problems, it can be very helpful for both

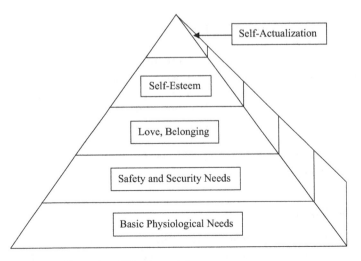

Figure 8.1 Maslow's Hierarchy of Needs model.

patients and therapists to use Maslow's hierarchy of needs for conceptualizing the relationships among their problems, needs, resources, and strengths.

Overall Complexity of Needs

In addition to prioritizing needs, it is important to consider the complexity of patients' problems as a whole. The presence of comorbid substance dependence, the presence of Axis II problems in addition to Axis I disorders, or having significant medical, family, or financial problems often presents a far more complex and challenging biopsychosocial situation for both the patient and therapist. Significant comorbidity within just the psychological domain is common, and having coexisting problems across the biopsychosocial domains occurs frequently as well (see Chapter 3).

At a basic level, the complexity of patient problems can be conceptualized as falling on a continuum that ranges from no to mild, moderate, and major complexity. Patients without clinically significant mental health problems or concerns would normally be assessed as having problems of essentially no complexity, while those with one or a small number of problems of lesser severity would be viewed as having problems of mild complexity. Patients with problems of serious complexity would typically include those with multiple problems of moderate or severe levels of need and/or risk (Table 8.9). Increased chronicity of problems also tends to increase problem complexity. There is no straightforward formula for assessing need complexity because the interaction of resources and needs across the many areas of peoples' lives results in a tremendously complex array of possibilities. Consequently, this assessment relies heavily on clinical judgment. Cases that typically involve more complex problems include serious persistent mental illness; mental illness or substance dependence that is treatment-resistant; serious and/or chronic employment, financial, legal, or relationship problems; serious cognitive disability; and significant comorbidity. On the other hand, it is not uncommon for

Table 8.9 Overall Complexity of Patient Needs

Level of Complexity	General Guideline
No or Very Little	Minimal or no clinically significant behavioral health problems or concerns; significant strengths prevent issues from developing into clinically significant problems
Mild	A small number of problems, usually of lesser severity; presence of strengths helps mitigate their effects
Moderate	Intermediate number of problems, usually of intermediate severity, and intermediate number of strengths
Major	Multiple problems of moderate or severe levels of severity and/or risk; strengths insufficient to counterbalance problems

individuals to have a serious problem in just a single area, and so while the complexity of their problems may be low, the seriousness of the needs or risks they face might be substantial nonetheless. For example, a college student who enjoys strengths and resources in many areas might experience serious destabilization surrounding the difficulty of a particular course or relationship problem.

Having more complex and serious behavioral health and biopsychosocial needs is often associated with more comprehensive and detailed psychological assessment. More sources of assessment data are typically needed in these cases to gain an adequate understanding of the case. For example, patients with major problems across the biopsychosocial domains may need medical and neuropsychological evaluations along with significant input from family members, employers, teachers, parole officers, or others. Thorough detailed assessments are common in inpatient psychiatric and substance abuse treatment programs, where patients often experience significant problem complexity.

Integrating Assessment Information

There appears to be little disagreement within the field regarding the importance of the above recommendations for the basic elements of psychological assessment. There is widespread consensus regarding the general purposes of assessment, the biopsychosocial domains and components that should be included, the need to consult multiple sources to obtain reliable and complete assessment information, as well as the need to evaluate the severity, complexity, and priority of the needs that patients face across the important areas of their lives.

There is less consensus in the field, however, regarding the further analysis and integration of the information collected during the assessment process. The topic of the integration of psychological assessment information has been examined at great length, though often within the perspective of particular theoretical orientations (e.g., within the psychoanalytic tradition).

The approach discussed thus far in this chapter has been largely descriptive and atheoretical. This is consistent with contemporary psychiatric assessment as reflected in the five-axial DSM diagnostic system. The biopsychosocial approach advocated in the present volume extends beyond a primarily descriptive approach, however, because it emphasizes *development* in addition to current functioning. Examining the development of personality characteristics and psychopathology and the etiology of mental health disorders goes well beyond the atheoretical description of current functioning. Questions of etiology are also critical to treatment planning because therapists often want to target treatment at the causes of individuals' problems so that the problems get resolved on a long-term basis as opposed to reducing symptoms and distress temporarily. Therefore, psychological assessment from a biopsychosocial approach tends to be comprehensive and explanatory and not merely descriptive.

Unfortunately, there is a limited amount of empirical research that helps inform the integration of psychological assessment information from a comprehensive, developmental, biopsychosocial perspective. Only two studies were located that

Table 8.10 Levels of Comprehensiveness and Integration of Psychological Assessments

Score	Rating Description
0	Assessment is missing critical biological, psychological, and sociocultural information in the context of the particular case.
1	The clinician obtained information regarding a variety of components across the biological, psychological, and sociocultural domains, but a lack of focus and attention to important concerns could lead to less effective treatment.
2	Basic competency. The clinician obtained comprehensive biological, psychological, and sociocultural information, and there is some evidence of integration of this information to address the patient's most important concerns.
3	The clinician obtained comprehensive biological, psychological, and sociocultural information; obtained information about some of the strengths and weaknesses the patient possesses; and the integration of this information helped prioritize the patient's concerns and problems.
4	The clinician addressed the patient's strengths and weaknesses comprehensively across the BPS domains, with attention given to individual and sociocultural differences. This information is integrated so that strengths will be reinforced and amplified and weaknesses and problems will be addressed. Issues are prioritized to reflect the patient's needs, circumstances, and preferences, and to maximize treatment effectiveness.

Source: Adapted from Meyer and Melchert (2011).

directly address this question. In the first of these, McClain, O'Sullivan, and Clardy (2004) investigated whether a sample of 79 psychiatric residents formulated integrative case conceptualizations according to a biopsychosocial framework. The study found that, on average, none of the groups of residents (first through fourth year, from four different institutions) wrote a biopsychosocial case formulation that reached what was identified as the basic level of competency. The reports typically included information regarding a wide range of biological, psychological, and sociocultural factors, but the information was not well integrated and consequently was judged to have the potential to lead to problems in treatment.

A second study, conducted by Meyer and Melchert (2011), found similar results. This study examined the treatment records for a sample of 163 therapy outpatients to evaluate the comprehensiveness and detail of the written assessment documentation and the extent to which that information was integrated and formulated in a manner that would maximize treatment effectiveness. Table 8.10 provides the rubric used to rate the level of comprehensiveness and integration of assessment information in that study.

The highest level in this rubric is where comprehensive information is considered in the context of patients' psychological development and biopsychosocial circumstances, with attention to both strengths and weaknesses, and where all these factors are prioritized and organized in a manner that maximizes the likelihood of treatment effectiveness. The mean rating of the files in the Meyer and Melchert

(2011) study was only 1.17 (SD = 0.45), and only 14.1% of the files were rated at 2 or higher on the scale (the midpoint indicating "basic competency"). Table 8.11 provides examples for each level of comprehensiveness and integration for three different types of patient cases.

The findings from the Meyer and Melchert (2011) and McClain et al. (2004) studies suggest that assessment information is too often reported descriptively, with too little depth and detail, and without an analysis and integration that explains the patient's current problems and concerns in the context of current biopsychosocial circumstances and developmental history. Patients' strengths and weaknesses need to be incorporated into assessments, and their needs have to be prioritized and considered in the context of their circumstances and developmental history so that individualized treatment plans are developed that maximize the likelihood of treatment effectiveness.

Conclusions

Conducting psychological assessment according to the guidelines discussed above is a complex process. Information regarding the large number of influences on individuals' lives needs to be gathered and then integrated in a manner that explains patients' development and current functioning and leads to the development of effective treatment plans. This comprehensive approach is needed to meet the basic purposes of psychological assessment noted at the beginning of this chapter. It allows for the identification of behavioral health problems and concerns that need clinical attention and provides the information needed for developing comprehensive case conceptualizations and treatment plans. It provides the baseline data for conducting ongoing assessment over the course of treatment to monitor progress and refine the treatment plan and refocus interventions as needed. The baseline data provided also allow for an outcomes assessment that can help measure the effectiveness of treatment.

As was emphasized earlier in this chapter, the approach one takes to psychological assessment varies significantly depending on the purposes. Across all types of purposes, however, the above guidelines apply. Psychologists need to be concerned with all the important areas of patients' lives across the biopsychosocial domains; they need to be attentive to the reliability and usefulness of assessment information; the information needs to be sufficient for the purpose; the severity and priority of patients' needs must be evaluated; and their overall complexity needs to be assessed as well. A comprehensive, holistic integration of the assessment data is then conducted so that the assessment can effectively inform the remaining phases of the treatment process.

Communicating to patients a thorough understanding of their behavioral health needs and biopsychosocial circumstances and then addressing them in ways that conform to their values, preferences, and sociocultural context are also important to developing therapeutic relationships and alliances with patients. These are

Table 8.11 Examples of Comprehensive and Integrated Psychological Assessments

Score	Anxiety Example	Depression Example	Adjustment Example
0	Patient presents with symptoms related to an anxiety disorder and indicates she has been under the treatment of a physician for these concerns for several months. She states he prescribed anxiolytics for these symptoms but she does not like taking the medication. Her symptoms have recently gotten more severe. In the intake assessment, the therapist does not obtain information related to medication, including what the patient is taking, side effects, efficacy, or medication adherence.	Patient presents with symptoms of depression, but the therapist does not obtain information related to suicidal ideation.	Patient presents with adjustment concerns related to her pending divorce, but the therapist does not obtain sociocultural information regarding the quality of her social support network.
1	Patient presents with symptoms related to an anxiety disorder and indicates she has been under the treatment of a physician for these concerns for several months. She states he prescribed anxiolytics for these symptoms but she does not like taking the medication. Her symptoms have recently gotten more severe. In the intake assessment, the therapist obtains some information regarding the medication the patient has been taking, but neglects crucial components such as side effects and medication adherence.	Patient presents with symptoms of depression, and the therapist obtains some information regarding psychiatric history, but neglects important components such as previous diagnoses of mood disorders, previous treatments and their outcome, and past psychological traumas.	Patient presents with adjustment concerns related to her pending divorce, and therapist obtains some sociocultural information regarding her relationships but does not assess the quality of her current relationship with her soon-to-be ex-husband.

(Continued)

Table 8.11 (Continued)

Score	Anxiety Example	Depression Example	Adjustment Example
2	Patient presents with symptoms related to an anxiety disorder and indicates she has been under the treatment of a physician for these concerns for several months. She states he prescribed anxiolytics for these symptoms but she does not like taking the medication. Her symptoms have recently gotten more severe. The therapist obtains information regarding the medication the patient has been taking, including important components such as side effects and medication adherence. The therapist notes that these issues may be related to current problems.	Patient presents with symptoms of depression, and the therapist obtains information related to psychiatric and personal history and past traumas; therapist shows link between current symptoms and this history.	Patient presents with adjustment concerns related to her pending divorce, and the therapist obtains important sociocultural information regarding her relationships, and notes how these relationships are stressful and beneficial to her.
3	Patient presents with symptoms related to an anxiety disorder and indicates she has been under the treatment of a physician for these concerns for several months. She states he prescribed anxiolytics for these symptoms but she does not like taking the medication. Her symptoms have recently gotten more severe. The therapist obtains information regarding the medication the patient has been taking, including important components such as side effects and medication adherence. The therapist addresses how the medication has helped reduce anxiety symptoms in recent months, and how side effects of the medication have had negative effects.	Patient presents with symptoms of depression, and the therapist obtains information related to personal history; therapist shows link between current symptoms and history; therapist obtains information from the patient to help understand detrimental and beneficial patterns of responses to life events.	Patient presents with adjustment concerns related to her pending divorce, and the therapist obtains information regarding her relationships and notes how these factors relate to her current concerns; therapist also obtains information regarding how marriage and divorce have been detrimental to her functioning and have had positive impacts as well.

4

Patient presents with symptoms related to an anxiety disorder and indicates she has been under the treatment of a physician for these concerns for several months. She states he prescribed anxiolytics for these symptoms but she does not like taking the medication. Her symptoms have recently gotten more severe. The therapist obtains information regarding the medication the patient has been taking, including important components such as side effects, medication adherence, and how the medication has helped ameliorate her anxiety symptoms in recent months, but also unpleasant side effects such as weight gain and tiredness. The therapist notes the concern that treatment of the anxiety with medication only has not actually helped manage the effects of her anxiety, but merely managed the symptoms. The therapist integrates what the patient has reported as advantages and disadvantages to taking the medication, as well as to develop an understanding of the case in terms of her own family of origin and developmental history.

Patient presents with symptoms of depression and therapist obtains information related to personal history; therapist shows link between current symptoms and history; therapist obtains information about how various factors are detrimental to the patient's functioning and how other factors have been helpful. Therapist makes links between patient's personal history and current functioning and takes into account patient concerns and personal preferences for treatment in order to maximize treatment effectiveness.

Patient presents with adjustment concerns related to her pending divorce, and the therapist obtains sociocultural and historical information regarding her relationships and notes how these factors relate to her current concerns; therapist obtains information regarding how marriage and divorce have been detrimental to her functioning and have had positive impacts as well. Therapist prioritizes the patient's problems and takes patient's preferences and needs into account in order to maximize treatment effectiveness.

Note: Adapted from table 3.6 in Meyer (2008).

necessary for addressing the third basic purpose of psychological assessment noted at the beginning of this chapter: to engage patients in the treatment process through a collaborative approach aimed at helping patients develop insight into the nature of their problems. There is strong evidence that the therapeutic relationship and alliance are among the strongest predictors of treatment outcome (see Chapter 10). Engaging patients in psychological assessment in this manner promotes the development of collaborative treatment relationships, which are associated with positive treatment outcomes.

There is, of course, much more to the behavioral health treatment process than just psychological assessment. After assessment information is appropriately collected, evaluated, and integrated, it is then used to develop a plan for treating patients' problems and concerns and building their resources and strengths. That is the topic of the next chapter. Before those issues are addressed, the following case example illustrates how psychological assessment can differ between a traditional approach and a biopsychosocial approach.

Case Example: A Cognitive-Behavioral Versus a Biopsychosocial Approach to Assessment with a Mildly Depressed Patient

Cognitive-Behavioral Approach to Assessment

A 49-year-old married Caucasian male presents with concerns about depressed mood. The patient first consults a psychologist with a cognitive-behavioral theoretical orientation. The psychologist reviews the intake questionnaire that the patient completed and notes that the patient wrote, "My wife wants me to see a psychologist for mild depression." On the checklist, the patient did not indicate concerns about his physical health, marriage, work, or finances. He indicated that he has two children who are both doing well. He also denied any suicidal ideation or disturbing thoughts or feelings and denied using alcohol or other drugs more than "socially."

After reviewing the questionnaire and quickly scoring the scales, the psychologist asks the patient to explain more about his depressed feelings. The patient reports that he is an attorney and has not been enjoying his work, his friends, or his family life the way he used to. He reports he may have a biological predisposition to depression because his mother was probably depressed when he was young. He reports that he appears to be well respected at work and his family is financially secure. He says that he is not arguing or fighting excessively with his wife or children, but "I am not really finding satisfaction in my family life, my work, my friends, or really anything else for that matter. Everything is getting to be a chore. I'm bored and I'm not particularly happy. I suppose I'm not that much fun to be around either, and my wife suggested that I see a psychologist. So here I am."

The patient goes on to explain that he has been married for 16 years to a physical therapist who enjoys a good reputation and a steady practice at a local orthopedic clinic. He reports having a good marriage and family life, though he feels that he and his wife slowly drifted apart after they began having children and became more settled in their careers. He says he was originally "head-over-heals infatuated with my wife. She was such an intelligent, funny, and positive person, and also incredibly fit and attractive." He says she is still all of those things and his friends and relatives "all think she's amazing." He and his wife have two children, a 16-year-old boy and a 19-year-old girl, who are both healthy and doing well—the oldest just finished her first year in college. He reports that he spends a lot of his nonwork time at home with his family and has attended many of the children's sporting and school events. Though he says he enjoys his family life, he also feels "like it's a chore. After 10 or 15 soccer games or piano recitals, it's not easy to get very excited about the next one." He reports having grown discouraged with his work because he feels he is not so much helping people find fair resolutions to their legal disputes, like he used to think, as much as just generating billable hours for the practice.

At this point, the psychologist explains that he thinks the patient is showing very typical signs of depression. The psychologists points out that his statements show that he engages in dichotomous thinking, as when he implied that if he isn't completely happy with his professional accomplishments, then he feels like they aren't satisfying or worthwhile at all, or that if he does not enjoy all his interactions with his wife or his children, then his marital and family life are not rewarding or meaningful at all. The psychologist also notes that he tends to overgeneralize from some of the less satisfying aspects of his work and child rearing and concludes in a blanket fashion that all of his work and child rearing responsibilities are not satisfying.

Biopsychosocial Approach to Assessment

The patient also consults a second psychologist who takes a biopsychosocial approach to treatment. The patient completes a similar intake questionnaire and the psychologist inquires about the same types of initial topics. The patient relates the same general information, though the psychologist asks for more details regarding many of the topics. For example, when the patient reports that he has good physical health and has had no major injuries or illnesses, the psychologist asks about his level of physical exercise and activity. The patient reports that he used to exercise regularly, almost as much as his wife, but he stopped over a decade ago. He reports eating too much, though he keeps his drinking to a minimum because "I know that that can get out of control too easily." He says he now weighs about 20 pounds more than he'd like to. The psychologist also asks for more detail regarding his marital relationship, and the patient reports that he and his wife gradually stopped going out on dates with each other after they had their second child—it cost a lot of money and it was too much trouble to find a good babysitter. He reports that his wife goes out with her girlfriends regularly, but he doesn't see friends often, in part because some of his best friends moved out of town several years ago.

The patient is asked about several additional topics, including hobbies and interests. He reports that he used to enjoy music, art, film, and theater, but that now he mostly just follows sports. He reports that he hasn't thought much about alternative work possibilities. About a decade ago, he thought about getting involved in the state bar association and serving on the board of a local nonprofit organization, but has since lost interest in those ideas. He reports that he has gradually drifted away from his own original family— he rarely calls or sees his parents or siblings except on holidays. He says that some of the personal emptiness he's feeling undoubtedly comes from his parents. They are both alive but have "a distant, perfunctory type of relationship with each other." He says they've always been that way—"Dad was always focused on work and Mom kept a perfect-appearing home. But they didn't show any signs that they really liked each other. They didn't really share that much in common." He then reports that "As my own kids are beginning to leave home, to be honest, I'm afraid that I'm becoming just like my own father.

He doesn't have any reason to get up in the morning. I could end up like him. What am I going to do that anybody cares about? On the one hand, I can't wait to retire because I'm getting so bored at work. But on the other hand, I'm sort of terrified that I won't have anything to do. I'll be a complete waste. Why would I even want to live?"

The psychologist explains to the patient that she thinks he is experiencing a pattern of worry and concern that is not uncommon for many individuals who grew up in a family like his. The psychologist shares her opinion that he has been highly successful in his life so far, at least outwardly—he has a highly respectable profession, a wife and children who are admired and successful, financial security, and good physical health. She explains that, despite all this success, he appears to have also succumbed to some of the same problems his parents had, which is not surprising. Though they apparently raised him to be very responsible and successful, a person who many others would be envious of, his parents appear not to have been particularly good at developing and maintaining intimate relationships with each other or their children. The psychologist explains that there are many different ways to address these issues, but before they decide on the best approach, she would like to hear his wife's perspective on these issues. She asks how much he has talked with his wife about these concerns and whether he would ask her to come in for their next appointment.

At the end of the discussion, the psychologist completes the following table (Table 8.12) with the patient to help summarize the information that the patient shared about himself and his situation. The patient and psychologist together discuss the level of need or strength in each area.

Table 8.12 Summary of the Biopsychosocial Assessment for the Case Example

Biopsychosocial Domains and Components	−3 Severe need	−2 Moderate need	−1 Mild need	0 No need	+1 Mild strength	+2 Moderate strength	+3 Major strength
Biological							
General physical health			X				
Childhood health history				X			
Medications				X			
Health habits and behaviors		X					
Psychological							
Level of psychological functioning		X					
History of present problem			X				
Individual psychological history			X				
Substance use and abuse			X				
Suicidal ideation and risk assessment			X				
Effects of developmental history			X				
Childhood abuse and neglect			X				
Other psychological traumas				X			
Mental status examination							X
Personality styles and characteristics					X		
Sociocultural							
Current relationships and social support			X				
Current living situation					X		
Family history			X				
Educational history							X
Employment						X	
Financial resources							X
Legal issues/crime				X			
Military history				X			
Activities of interest/hobbies			X				
Religion				X			
Spirituality				X			
Multicultural issues				X			

9 Treatment Planning

Receiving a psychological diagnosis and assessment can bring relief to individuals who have been suffering with complicated mental health issues. After spending months or years in psychological confusion and pain, a diagnosis can provide critical reassurance that one's existence is perhaps not spinning out of control but is actually part of an predictable and understandable pattern, even if it is an agonizing and dysfunctional pattern. Patients receiving a medical diagnosis in analogous circumstances often experience a similar type of relief.

Patients typically want much more than just a diagnosis, however. Even if some comfort is gained from knowing that others have similar kinds of problems and that researchers have identified a syndrome that helps explain the etiology and phenomenology of the disorder, attention generally turns quickly to how those problems can be ameliorated and resolved. At this point, the assessment phase of treatment evolves into the treatment planning phase.

In traditional approaches to treatment planning, therapists often formulate explanations of patients' personality and psychopathology from a particular theoretical orientation and then implement the treatment approach that is indicated by that orientation (Garb, 1998). If a cognitive therapist conceptualized a patient case involving major depression, for example, he or she was likely to identify depressogenic thoughts and beliefs that were causing the depression and then would identify a plan for replacing those thoughts with rational ones. If a biologically oriented psychiatrist was referred that same patient, he or she was likely to include antidepressant medications in the treatment plan.

The biopsychosocial approach to treatment planning rests on a very different approach. This approach uses a holistic perspective that focuses on the full range of psychological, biological, and sociocultural influences on development and functioning along with their interactions. Emphasis is placed on achieving positive health and functioning across the important areas of patients' lives in addition to relieving psychological distress and reducing symptomatology. Psychologists working from this perspective employ a range of evidence-based therapies and other strategies to build strengths and assets at the same time as they treat problems, taking advantage of the synergy that is possible when strengths are bolstered while problems are simultaneously lessened across the biopsychosocial domains.

This chapter examines the basic conceptual issues and processes that form the foundation for approaching behavioral health treatment planning from a

Foundations of Professional Psychology. DOI: 10.1016/B978-0-12-385079-9.00009-6

biopsychosocial perspective. Before beginning this examination, however, an important preliminary issue needs to be addressed first. The biopsychosocial approach to professional psychology is based on a health care orientation to clinical practice. As a result, treatment planning from this perspective needs to begin with a consideration of the safety and effectiveness of the interventions that might be used with patients. Rather than focus on what the field has to offer in terms of therapies that have been developed, the focus needs to be on offering therapies and interventions that are safe and effective. Therefore, this chapter begins with a discussion of this basic question before moving on to outline the process of treatment planning from a biopsychosocial approach.

A Critical Preliminary Issue

When physical or mental health patients seek assistance for medical or psychological problems they are experiencing, health care professionals need to be confident that their assessment findings are accurate and the interventions they recommend for resolving those problems are safe and effective. Across health care fields, professionals need to ensure that their practice is supported by reliable empirical evidence regarding the safety and effectiveness of the treatments they recommend and provide to patients (American Psychiatric Association, 2006; Institute of Medicine, 2001). The ethical principles of nonmaleficence and beneficence obligate health care professionals to provide interventions that are beneficial and do not introduce disproportionate risks of unwanted side effects or harm. The credibility of psychology as a health care profession depends on the ability to provide care that is consistent with these principles.

Fortunately, there is a very solid support regarding the effectiveness of a wide range of psychological interventions. With regard to mental health treatment in general, there is widespread agreement, as Norcross, Beutler, and Levant (2006) put it, that "most mental health treatments have already been empirically established as effective and as safe as other health care and educational interventions" (p. 404). As was noted in Chapter 4, consensus regarding this conclusion has been growing ever since Smith and Glass (1977) published their landmark meta-analyses in the late 1970s.

Several more specific conclusions supporting the use of psychotherapeutic interventions are also well supported by research evidence. These were summarized by Reed and Eisman (2006) as follows (see also Lambert & Archer, 2006; Norcross et al., 2006; Wampold, 2001):

1. Psychotherapy is generally effective, with positive outcomes reported for a wide variety of theoretical orientations and treatment techniques.
2. Although there is some variability across disorder, the effects of psychotherapy are generally as good as or superior to the effects of psychotropic medications for all but the most severely disturbed patients.

3. The outcomes of psychotherapy are substantial across a variety of relevant areas, including psychiatric symptoms, interpersonal functioning, social role performance, and occupational functioning.
4. Psychotherapy is relatively efficient in producing its effects.
5. The outcomes of psychotherapy are likely to be maintained over time, particularly in contrast to the effects of psychotropic medications . . .
6. Psychotherapy may offset the costs of medical services by reducing hospital stays and other medical expenses (pp. 16–17).

The widespread agreement regarding the above issues allows psychologists to proceed with providing behavioral health care in an ethically and professionally responsible manner. There are a number of unresolved issues in the therapy outcome literature, however, that require psychologists to proceed cautiously. The most controversial of these questions concerns the possible differential effectiveness of several specific therapies. The effectiveness of therapy is discussed in more detail in the next chapter on treatment, but the issue is briefly reviewed here to establish the justification for proceeding with treatment planning.

Over the history of the field, many psychologists believed that particular psychotherapy approaches were best suited for certain types of psychological issues (e.g., behavior therapy for anxiety, cognitive therapy for depression, psychodynamic therapy for character issues), and extensive outcome research appeared to support this conclusion. In the 1990s, Division 12 (Clinical Psychology) of the APA formed a task force to examine this question, and their report supported the view that particular treatments were effective for treating certain disorders, while other treatments could not be relied on for producing positive outcomes for those, or perhaps any, disorders (Chambless & Hollon, 1998).

At the same time, however, there is substantial evidence regarding the importance of common factors for explaining the effectiveness of psychotherapy. Wampold (2001), for example, found that the effect of the particular treatment employed accounts for only a negligible amount of variance in therapy outcome, whereas the effect of the quality of the therapist is of major importance in explaining outcomes. Wampold concluded that the effect of therapist abilities on treatment outcome was at least 10 times the effect of differences among treatments. Therapists' abilities to develop a therapeutic relationship and develop a shared understanding with the patient regarding how to proceed with treatment (i.e., the therapeutic alliance) were found to be critical factors in determining the outcomes of therapy, whereas the type of therapy that was implemented mattered little in explaining therapy outcomes. These findings tend to support the famous conclusion reached by the Dodo bird in *Alice in Wonderland* that "Everybody has won, and all must have prizes." This *Dodo Bird Verdict* was reached by Rosenzweig already in 1936 and has been supported by many meta-analyses since.

These highly divergent perspectives regarding the effectiveness of psychotherapy point to two general schools of thought regarding the best approach to evidence-based practice within professional psychology. One school advocates that the best approach to offering behavioral health care begins with the diagnosis of a mental disorder, which is then followed by the selection and application of an evidence-based intervention to treat that disorder. On the other side are those who focus less on the diagnosis and more on the particular needs and characteristics of the individual patient. The emphasis here is on establishing a strong therapeutic relationship and developing a treatment plan tailored specifically to the personal characteristics, circumstances, and history of the patient rather than focusing primarily on his or her diagnosis (Davidson, 2000; Goodheart & Kazdin, 2006; Messer, 2004; Sobell & Sobell, 2000). Goodheart and Kazdin (2006) noted that many, and perhaps most, psychologists do not practice at either of these extremes but rather combine elements of both. Nonetheless, there is a significant divide between these two camps regarding the main implications of the psychotherapy outcome research (e.g., see Ollendick & King, 2006; Wampold, 2006).

Research is emerging that will likely resolve these diverging views on the effectiveness of psychotherapy, though it may take some time before consensus is reached (see Chapter 13). At this point, the evidence clearly indicates that there are many effective treatments for many types of disorders and patients, and that the therapeutic skills of the individual therapist are critical to explaining therapy outcomes. There are important implications of these findings for treatment and outcomes assessment that are examined in detail in the next two chapters. And therapists clearly also need to stay current with the research that investigates these issues. For purposes of proceeding with treatment planning, however, the general conclusions discussed at the beginning of this section are supported by extensive replicated research, and psychologists can be confident in recommending a variety of effective psychological treatments to their patients.

In addition to the question of treatment effectiveness, however, is the question of the safety of treatments. In Chapter 6 (in the section on nonmaleficence), it was noted that there have been growing concerns within the field regarding risks of harm that may be imposed by psychological treatment. This issue gained widespread attention in the 1990s when controversies regarding repressed memories of child abuse grew highly contentious. Other recent therapies for which there is evidence of potential or actual harm include rebirthing attachment therapy (Chaffin et al., 2006), group interventions for antisocial youth (Weiss et al., 2005), conversion therapy for gay and lesbian patients (APA, 2000b), critical incident stress debriefing (Mayou, Ehlers, & Hobbs, 2000), and grief therapy (Bonnano & Lilienfeld, 2008).

It is critical that psychologists keep current with the research regarding potentially harmful psychological treatments in addition to the literature regarding effective treatment responsible clinical practice and to ensure that clinicians are meeting their obligations of beneficence and nonmaleficence. This issue is also addressed in the next two chapters in terms of processes and techniques that help ensure that patient lack of progress is monitored and the effectiveness of treatment is maximized. When these issues are all taken into consideration, psychologists can proceed with treatment planning based on solid evidence that there are a variety of safe and effective psychotherapies available for treating a wide range of behavioral health problems.

Treatment Planning from a Biopsychosocial Perspective

A biopsychosocial approach to treatment planning focuses on meeting patients' behavioral health needs and promoting their biopsychosocial functioning from a comprehensive holistic perspective. After an integrative, holistic evaluation of the patients' needs is conducted, a plan is developed to address those needs within the context of the individual's unique developmental history and current circumstances and in a manner designed to maximize treatment effectiveness. Sometimes there are critical or emergency needs that require immediate attention (e.g., suicidality, the well-being of the children of an unstable parent). At other times, the gradual process of building social and interpersonal skills, examining dysfunctional personality characteristics, or addressing existential questions unfolds over an evolving long-term therapy relationship. Sometimes therapy is delayed for the time being because, for example, certain issues need to be addressed (e.g., substance abuse or employment problems) or resources need to be strengthened (e.g., personal coping resources or external social supports) before it is prudent to examine particularly difficult or stressful therapy issues.

Treatment planning from a biopsychosocial perspective is consequently a complicated process. In traditional approaches to treatment planning, therapists often recommended a treatment approach based on their adherence to a particular theoretical orientation. A biopsychosocial approach, on the other hand, requires an individualized evaluation of patients' needs and circumstances across the full range of biopsychosocial areas. Those needs then need to be prioritized with the aim of maximizing treatment effectiveness and preventing harm. Treatments that have been shown to be safe and effective for applying in particular cases (e.g., given the patient's particular developmental history and biopsychosocial circumstances) can then be used to address issues and concerns, often with the aim of realizing the synergy that can result when problems and deficits in certain areas are addressed while relying on and further developing strengths in other areas.

This complicated process begins with an elementary decision, however. Before recommending a treatment plan, the therapist needs to evaluate whether intervention is the appropriate way to proceed.

Starting at the Beginning: Deciding Whether to Intervene

As emphasized in the previous chapter, patients present with a remarkable diversity of issues, ranging from mild isolated problems within the context of substantial strengths and resources to highly serious and complex coexisting problems across numerous biopsychosocial areas. Interventions for addressing these many different situations vary greatly as well. Rather than assume that treatment is indicated whenever a problem or concern is identified, psychologists first need to evaluate whether intervention is the best way to address the problems or concerns that patients present with. This evaluation tends to revolve around the severity and complexity of the issues that were identified in the psychological assessment and normally includes just four options:

1. *Do not intervene* because the problem or concern does not warrant clinical intervention.
2. *Intervene* with clinically significant concerns and problems.
3. *Postpone a decision about intervening*, but observe and monitor the problem in the meantime ("watchful waiting").
4. *Refer* to another professional for more assessment or to provide the needed intervention.

Patients are sometimes concerned about issues that do not rise to the level of clinical concern in terms of significant impacts on their behavioral health or biopsychosocial functioning. Upon learning that their concerns or experiences are common and/or have no significant implications for their health and functioning, many patients can be reassured that no further assessment or intervention is needed (e.g., that a particular emotional, cognitive, behavioral, or sexual response that they experience is not unhealthy). Of course, these patients should feel free to return to their behavioral health care provider if these or other issues become concerning at a later point.

When a patient has a clinically significant problem and a psychologist has sufficient clinical training and experience for treating it within the context of the patient's background and life circumstances, then the psychologist often provides the services needed. Sometimes a decision not to provide the needed intervention is based on factors that have very little to do with the patient (e.g., the psychologist currently has too many suicidal patients to responsibly take on another one). Usually, however, the decision is made on the basis of the psychologist's competencies. If one has not had sufficient education and clinical experience to safely and effectively provide the needed intervention, it is normally appropriate to refer the patient to another therapist.

When it is unclear whether a patient has a clinically significant problem or concern or there is reason to believe that a problem might resolve on its own, there is a middle option between providing and not providing an intervention that is often very useful. In these cases, it may be appropriate to postpone intervention and instead observe and monitor the patient over time to evaluate the progression of the issues. "Watchful waiting" is common in medicine because some situations or processes are near the boundary of clinical significance and/or there is a reasonable likelihood that they may resolve on their own with minimal or no intervention. It is important in both physical and behavioral health care to intervene early to prevent problems from developing or increasing in severity. But it is also important not to waste resources or risk undesirable side effects if there is a reasonable likelihood that a problem will improve on its own or it is unclear whether a clinically significant problem even exists. This approach can also be useful when very limited interventions (e.g., psychoeducation) are implemented following an inquiry about a problem of mild severity and complexity.

Referring patients to other professionals is often done because one does not have sufficient expertise to provide the needed intervention. It is not uncommon for psychologists to possess the experience and expertise to diagnose a problem (e.g., substance dependence, relationship conflict, and family dysfunction) but not the expertise for providing the needed treatment (e.g., substance abuse treatment or family therapy). Consequently, psychologists commonly work collaboratively with a variety of professionals to meet patients' needs. For example, psychologists frequently collaborate with other psychologists, physicians, psychiatrists, law enforcement officials, educators, social workers, and others to meet particular patients' needs.

Addressing Severity and Complexity of Need

Decisions involving whether, when, and how to intervene are normally based on the severity and complexity of the needs that were identified in the psychological assessment. For example, when severe biopsychosocial needs are identified, intensive and immediate interventions are often indicated. Cases that require this type of response in behavioral health care frequently involve suicidality, homicidality, or other risks of injury to self or others (e.g., the abuse or neglect of children or vulnerable adults). When these issues are present, interventions addressing other issues often receive less attention or are postponed altogether until the emergency needs are satisfactorily addressed. When immediate, intensive care is needed, the target for achieving the most urgent treatment goals may be just hours or days, whereas the target for achieving less urgent medium-term goals may be several weeks. Still other goals for assisting a patient to return to or achieve stable biopsychosocial functioning may become part of the long-term plan. In the case of serious alcoholism, for example, immediate treatment for detoxification and medical stabilization may be needed acutely. Following that, relatively intensive substance abuse treatment, family therapy, and vocational counseling may be needed to address various medium-term goals. Over the long term, a mutual support group such as Alcoholics Anonymous may be used to maintain long-term sobriety.

Treatment planning, therefore, builds directly on the results of the assessment that was conducted. The complexity of patients' needs taken as a whole must be considered along with the severity of their individual needs in order to proceed with this process. Though this evaluation becomes quite complex in complicated cases, Table 9.1 provides basic options that are typically considered in relation to the level of severity of patient needs within individual biopsychosocial areas.

The interaction of patients' needs across the biopsychosocial areas and the complexity of their needs taken together as a whole can result in quite complicated treatment planning. Though extensive training and clinical experience are needed to gain the knowledge and skills needed to develop treatment plans for these types of cases, the remainder of this chapter reviews fundamental issues that inform treatment planning in all types of cases. This discussion starts with the decision-making models that have been developed to help identify the level and type of care that are

Table 9.1 Typical Treatment Planning Possibilities Associated With Level of Severity of Need for Individual Areas of Biopsychosocial Functioning

Severity of Need	Typical Treatment Planning Possibilities
+1 to an area of strength +3 or resource	• Reinforce strengths, amplify assets and resources (internal as well as external/environmental).
0 No evidence of need	• Do not intervene. • Build this area into a source of strength or a resource for the patient. • Refer patient back to referral source with opinion that no significant problem exists, and provide suggestions for future monitoring or prevention.
−1 Mild need	• Provide support, psychoeducation, and/or treatment for making changes. • Observe and monitor the problem or concern, make decision about intervening at later point. • Postpone interventions until higher-priority needs are addressed.
−2 Moderate need	• Provide intervention oneself. • Refer to other professional(s) to provide intervention. • Collaborate with other professional(s) on providing intervention(s). • Postpone intervention until higher-priority needs are addressed.
−3 Severe need	• Intensive and/or immediate interventions are often needed; monitor with extra care. • Provide intervention oneself. • Refer to other professional(s) to provide intervention. • Collaborate with other professional(s) on providing intervention(s). • Plan ongoing care, aftercare, and follow-up as needed.

appropriate for addressing patient needs at varying levels of severity and complexity.

Level of Care Decision Making

The level of care needed to reach treatment goals varies widely across patients. Clients with serious and persistent mental health conditions, for example, often require comprehensive multidisciplinary treatment to achieve stabilization and rehabilitation. These cases often require psychologists to collaborate with psychiatrists, nonpsychiatric physicians, nurses, rehabilitation counselors, occupational therapists, social workers, legal professionals, along with the patient's family members and others who together assess the level and types of care needed to assist the patient in achieving the highest levels of functioning and quality of life possible. At the other end of the continuum are patients with very limited and circumscribed problems who may need just a few therapy sessions to resolve their issues.

The need to identify the appropriate level of care grew quickly in the 1960s when mental health reforms resulted in the deinstitutionalization of the chronically mentally ill. Major outcomes of that movement involved providing care in the least restrictive manner possible along with greater involvement by the patients and their families in managing the care (Durbin, Goering, Cochrane, MacFarlane, & Sheldon, 2004). The legal requirement to provide treatment that restricts a patient's liberty the least while still remaining efficient and effective (*Project Release v. Prevost*, 1983) is based on the moral obligation to respect patients' rights to autonomy and not restrict autonomy without just cause. Clinicians also need to minimize the wasteful use of resources (which, after all, could otherwise be used to help others in need). On the other hand, clinicians also need to maximize the chances of positive outcomes by providing enough treatment to ensure a positive response.

Several approaches for assessing the level of care needed by patients have been developed (e.g., Anderson & Lyons, 2001; Durbin et al., 2004; Srebnik et al., 2002). The widely used system developed by the American Association of Community Psychiatrists illustrates several important components that are common in level-of-care models. The Level of Care Utilization System for Psychiatric and Addiction Services (LOCUS) was developed to match patients' behavioral health needs with the appropriate intensity of service and level of care needed to address and manage those needs (Sowers, George, & Thompson, 1999). The instrument includes six assessment scales that measure (1) level of risk of harm to self or others; (2) level of general functioning (e.g., ability for self-care, appropriate inter-action with others); (3) medical, addictive, and psychiatric comorbidity; (4) level of stress and level of support in the patient's environment; (5) the patient's treatment and recovery history; and (6) the patient's level of acceptance of responsibility for maintaining his or her health and engagement with helping resources. Scores on these scales are summed to help establish treatment recommendations. A modified version of the LOCUS has been developed for use with children and adolescents (i.e., the CALOCUS: Sowers, Pumariega, Huffine, & Fallon, 2003). It follows the same format as the LOCUS but incorporates considerations that are relevant for children (e.g., the sixth assessment dimension focuses on the primary caretaker's acceptance and engagement as well as the child's).

To assist individuals at all levels of need, the LOCUS and CALOCUS both identify six levels of care that represent increasingly intensive (and expensive) use of services. These six levels are (1) recovery maintenance and health management personally managed by the patient; (2) outpatient services; (3) intensive outpatient services; (4) intensive integrated service without 24-hour psychiatric monitoring; (5) nonsecure 24-hour services with psychiatric monitoring; and (6) secure 24-hour services with psychiatric management.

Graduated, "Stepped" Models of Intervention

The above level-of-care models were developed primarily to address the needs of those with serious chronic mental illness. A variety of other graduated level-of-care models have been developed to address particular mental health issues or the behavioral health needs of the population in general. For example, one well-known example developed to assist with concerns specific to sexuality is the PLISSIT model developed by Annon in 1976 (which is an acronym for Permission, Limited Information, Specific Suggestions, and Intensive Treatment). The model suggests providing patients with assurance or limited answers for common, less complicated concerns about one's sexuality and proceeds to more complex and specialized interventions for more complicated issues.

A critical area where a graduated level-of-care approach is necessary for deciding how to intervene involves suicidality. More intensive and immediate interventions for managing and treating suicide risk are implemented as the risk for suicide increases. For example, Rudd (2006) recommends a five-level assessment of suicide risk (i.e., none, mild, moderate, severe, and extreme). For patients at no or mild level of risk, no particular changes in ongoing treatment are recommended, though suicidal ideation is monitored on an ongoing basis. For those at a moderate risk level, increases in the frequency or duration of outpatient visits may be recommended. The need for inpatient hospitalization must be evaluated immediately for those at higher levels of risk.

Given the high prevalence of substance abuse in the Unites States and the substantial economic cost of associated medical, mental, and social problems, the US Substance Abuse and Mental Health Services Administration (SAMHSA) initiated the SBIRT program to expand treatment capacity and early intervention for substance abuse. This program was recently implemented in a variety of inpatient and outpatient medical settings to provide early intervention for those who are not dependent on substances but who may be engaging in problematic substance use or abuse (Clay, 2009). The SBIRT acronym is short for Screen, Brief Intervention, Brief Treatment, and Referral to Treatment. In this system, brief substance abuse screening instruments are incorporated into routine medical practice. If a moderate risk for substance abuse is indicated when patients complete a screening instrument, brief interventions are provided to try to increase awareness of substance use patterns and consequences and to motivate behavior change to reduce substance use. If moderate to high risks are identified, then more comprehensive brief treatment is provided. If severe risk or dependence are indicated, a referral for more extensive treatment is then provided.

The United Kingdom has undertaken a pioneering effort to implement a comprehensive graduated level-of-care model on a countrywide basis. In 2007, the Department of Health instituted the National Institute for Health and Clinical Excellence (NICE) treatment guidelines to improve the availability of evidence-based psychological treatments throughout the United Kingdom (Clark et al., 2009). The results of the initial psychological assessment determine the level of care that is provided. A wide range of interventions is included in the system, including self-help activities, computer-assisted therapy, or individual psychotherapy. Preliminary data on the effectiveness of this initiative are promising (e.g., effect sizes of $0.98-1.26$ have been found across a range of outcome measures; see Clark et al., 2009).

Level-of-care decision making is critical to treatment planning from a biopsychosocial perspective. When the severity and complexity of the full range of patients' needs are considered together as a whole, a treatment plan needs to be developed that addresses the severity of need in particular areas and the severity and complexity of all the patients' needs as a whole, as well as maximizes the likelihood of effectiveness over the long term. This approach is much more complicated than conceptualizing cases according to the perspective of a preferred theoretical orientation, but it is necessary when taking a health care orientation to behavioral health.

Collaborative Care

As health care has become more specialized, there has been an increase in the use of teams of health professionals to assess and evaluate cases, plan the treatment, implement the treatment, and monitor patients' health problems. In 1978, the World Health Organization began emphasizing the importance of teamwork and collaborative care, while more recently the Institute of Medicine (2000, 2001) has recommended collaborative care to improve the quality and effectiveness of health care in the United States. This approach is particularly important when treating more complex cases. Though collaborative care is widely assumed to be beneficial, most research to date has found only limited effects on patient outcomes or on costs and resource utilization (Bosch et al., 2010). Nonetheless, coordinated treatment plans that involve collaboration among multiple human service professionals are unavoidable when treating complex and serious cases. They also follow logically when applying a biopsychosocial perspective that addresses needs and functioning broadly across all areas of patients' lives.

Contextual Factors

Treatment plans ideally are developed through a collaborative process involving the patient, therapist, and other relevant stakeholders in patients' lives (e.g., family members, medical providers) so that patient motivation and internal and external supports are maximized. This ideal is often not achieved, however, due to a variety of contextual factors specific to each individual case. Some of these factors are internal to the patient while others are external, but they greatly complicate treatment in many cases. Consider the following examples:

- A suicidal adult refuses mental health evaluation, though family members are certain he represents a risk to himself or others.
- A husband wants to figure out how to reduce his wife's "nagging" and save his deteriorating marriage, but does not consider his drinking to be a problem.
- A teenager wants to try to get her parents to understand how much she loves a particular boy, while her parents believe that the relationship needs to end because their daughter may run away, become pregnant, and/or drop out of school.
- A compulsive shopper does not have insurance coverage and is unable to pay for treatment.
- A patient asks for assistance with eliminating his or her homosexual feelings because of the prohibitions of his or her family's religion or culture.

Factors such as these can have a major impact on treatment and consequently need to be incorporated into patients' treatment plans. There are a wide variety of such factors, as illustrated in Table 9.2. Identifying and effectively managing these

Table 9.2 Common Contextual Factors Affecting Treatment

Client's insight into own problems, acceptance of responsibility for own problems and own recovery	Client's coping style and related personality characteristics (e.g., resilience, impulsivity)
Client's level of family, peer, and other support	Lack of finances, insurance restrictions, transportation or geographic barriers
Client's level of stress	Cultural factors
Stability of patient's psychosocial environment	Disagreement among stakeholders in a client's treatment
Co-occurring medical, psychiatric, and substance use disorders	Legal or administrative issues affecting treatment (e.g., involuntary hospitalization, mandated treatment, evaluation for disability benefits)
Client's level of risk to self or others	Client's decision-making capacity is questionable
Strength of the therapeutic alliance	Client's treatment history and previous attempts to solve problems

factors requires substantial clinical knowledge and skill, but the success of treatment is often dependent on anticipating and working through and around these issues.

Ongoing Care and Follow-Up

Many patients are unable to maintain their treatment gains without ongoing monitoring, support, and care. This is especially important in cases of severe need and chronic conditions, but can also be important in cases of mild and moderate need. Ongoing care and follow-up interventions are critical considerations in cases of risk to self or other (e.g., Joiner, 2005; Rudd, 2006) and are also essential in the treatment and management of chronic conditions in general. When behavioral health services are focused on patients' health and biopsychosocial needs, follow-up and plans for ongoing care become routine considerations in treatment planning. The increased recognition of the importance of ongoing care is also reflected in the shift of the substance abuse treatment field from an acute to an ongoing care model (Hazelton, 2011).

Range of Alternative Interventions

Choosing the interventions that are most likely to be effective with particular patients with particular mental health issues and biopsychosocial circumstances is a very complicated topic that cannot be addressed within the scope of this volume. (Indeed, an entire graduate curriculum is needed to adequately address this topic.) The question of the number of therapies and types of interventions with which a single psychologist can be competent also cannot be addressed here. Nonetheless, to illustrate the wide variety of options that are available, Table 9.3 provides

Table 9.3 Examples of Possible Interventions Across the Biopsychosocial Domains and Levels of Severity or Need

Domains and Components	Strengths (+1 to +3)	No Problem (0)	Mild Severity (1)	Moderate Severity (2)	Severe (3)
Biological					
General medical history	• Reinforce healthy eating, exercise, and lifestyle	• Recommend more attention to diet and activity	• Persuade client to engage in healthier lifestyle • Help monitor compliance with prescribed treatments	• Refer for a physical exam • Help monitor compliance with prescribed treatments	• Refer for immediate physical evaluation or emergency care • Help monitor compliance with prescribed treatments
Childhood health history	• Reinforce healthy coping, adjustment, and treatment adherence if problems were overcome	• Reinforce healthy coping, adjustment, and treatment adherence if problems were overcome	• Support treatment adherence and healthy lifestyle if patient is still dealing with long-standing issues	• Support treatment adherence and healthy lifestyle if patient is still dealing with long-standing issues	• Refer for immediate physical evaluation or emergency care if needed • Help monitor compliance with prescribed treatments
Medications	• Reinforce healthy habits and healthy use of medicines and substances	• Reinforce healthy habits and healthy use of medicines and substances	• Refer for medical evaluation • Help monitor effectiveness and side effects of medications	• Refer for psychiatric evaluation • Help monitor effectiveness and side effects of medications • Coordinate family members to supervise patient medicine use	• Refer for immediate psychiatric evaluation or hospitalization • Coordinate family members to supervise patient medicine use
Health habits and behaviors	• Reinforce healthy eating, exercise, and lifestyle	• Recommend more attention to diet and activity	• Persuade patient to engage in healthier lifestyle	• Refer for a physical exam • Help patient develop healthy lifestyle • Help monitor compliance with prescribed treatments	• Refer for immediate physical evaluation or emergency care • Help monitor compliance with prescribed treatments

(Continued)

Table 9.3 (Continued)

Domains and Components	Strengths (+1 to +3)	No Problem (0)	Mild Severity (1)	Moderate Severity (2)	Severe (3)
Psychological					
Level of psychological functioning	• Reinforce positive mental health and role functioning • Amplify resources & strengths where helpful	• Reinforce positive mental health and role functioning • Recommend helpful psychoeducational interventions • Develop areas of strength and resource	• Individual therapy • Group therapy • Support group • Bibliotherapy • Internet resources • Develop strengths and resources	• More intensive treatment • Refer for evaluation for psychotropic medicine • Develop compensating strengths and resources	• Evaluate need for hospitalization • Increase frequency of individual therapy • Refer for psychiatric evaluation • Develop compensating strengths and resources
History of present illness/problem	• Reinforce positive mental health and role functioning • Amplify resources & strengths where helpful	• Reinforce positive mental health and role functioning • Recommend helpful psychoeducational interventions • Develop areas of strength and resource	• Individual therapy • Group therapy • Support group • Bibliotherapy • Develop strengths and resources	• Refer for evaluation for psychotropic medicine • Develop motivation to change • Develop compensating strengths and resources	• Evaluate need for hospitalization • Refer for psychiatric evaluation • Develop compensating strengths and resources
Individual psychological history	• Reinforce positive mental health and role functioning • Amplify resources and strengths where helpful	• Reinforce positive mental health and role functioning • Recommend helpful psychoeducational interventions • Further develop strengths and resources	• Conduct neuropsychological screening • Individual therapy • Group therapy • Support group • Bibliotherapy • Further develop strengths and resources	• Refer for neuropsychological evaluation • Consider intensive long-term therapy (e.g., psychodynamic, Acceptance and Commitment Therapy (ACT), Dialectical Behavior Therapy (DBT)) • Develop compensating strengths and resources	• Refer for neuropsychological evaluation • Consider residential or intensive outpatient treatment • Refer patient for vocational rehab., social services, etc. as needed

Substance use and abuse	• Reinforce positive mental health and role functioning • Amplify resources & strengths where helpful	• Reinforce positive mental health and role functioning • Recommend helpful psychoeducational interventions	• Discuss substance use and its consequences • Work on reducing substance use • Engage patient in mutual support group	• Refer for brief substance abuse treatment • Engage patient in mutual support group	• Refer family members for support group • Consider detox hospitalization • Refer for intensive substance abuse treatment
Suicidal ideation and risk assessment	• Reinforce positive mental health and role functioning	• Reinforce positive mental health and role functioning	• Outpatient therapy • Ongoing monitoring of suicidality	• More intensive outpatient therapy • Ongoing monitoring of suicidality	• Evaluate need for hospitalization • Refer to specialist
Individual developmental history	• Reinforce positive mental health and role functioning • Amplify resources and strengths where helpful	• Reinforce positive mental health and role functioning • Recommend helpful psychoeducational interventions	• Reinforce positive mental health and role functioning • Reinforce and further develop strengths and resources	• Reinforce positive mental health and role functioning • Reinforce and further develop strengths and resources • Refer to mutual support group	• Reinforce positive mental health and role functioning • Reinforce and further develop strengths and resources • Refer to mutual support group
Childhood abuse history	• Reinforce positive mental health and role functioning • Amplify resources and strengths where helpful	• Reinforce positive mental health and role functioning • Recommend helpful psychoeducational interventions	• Individual therapy • Bibliotherapy	• Individual therapy • Group therapy • Family therapy if appropriate	• Consider intensive long-term therapy (e.g., psychodynamic, ACT, DBT)
Other psychological traumas	• Reinforce positive mental health and role functioning • Amplify resources & strengths where helpful	• Reinforce positive mental health and role functioning • Recommend helpful psychoeducational interventions	• Individual therapy • Bibliotherapy • Further develop strengths and resources	• Consider exposure and related therapies	• Consider intensive long-term therapy (e.g., psychodynamic, ACT, DBT)

(Continued)

Table 9.3 (Continued)

Domains and Components	Strengths (+1 to +3)	No Problem (0)	Mild Severity (1)	Moderate Severity (2)	Severe (3)
Mental status examination	• Reinforce positive mental health and role functioning • Amplify resources and strengths where helpful	• Reinforce positive mental health and role functioning • Recommend psychoeducational interventions if they would be helpful	• Conduct neuropsychological screening	• Conduct psychological testing • Refer for neuropsychological exam	• Refer for psychiatric or neuropsychological exam
Personality style and characteristics	• Reinforce positive mental health and role functioning • Amplify resources & strengths where helpful	• Reinforce positive mental health and role functioning • Recommend helpful psychoeducational interventions	• Social skills training groups • Bibliotherapy • Recommend helpful psychoeducational interventions	• Individual therapy • Group therapy	• Consider intensive long-term therapy (e.g., psychodynamic, ACT, DBT)
Sociocultural					
Relationships and social support	• Reinforce healthy relationships	• Further develop strengths and resources	• Couple session to assess nature and severity of relationship issues	• Couple therapy • Communication skills training	• Develop safety plan if relevant • Refer patient to anger management program
Family history	• Maintain positive family relationships	• Further develop strengths and resources	• Family session to assess nature of family issues • Bibliotherapy	• Family therapy • Support group	• Family therapy
Current living situation	• Reinforce positive living situation	• Further develop strengths and resources	• Invite in roommates to conduct assessment session	• Facilitate temporary move to family member or friend	• Refer for evaluation of need for shelter and social services
Educational history	• Anticipate future educational and training needs	• Psychoeducation	• Refer for GED program, vocational training, job retraining, career counseling	• Complete psychological testing to determine nature and extent of cognitive deficits	• Refer for neuropsychology evaluation of cognitive deficits

Employment	• Maintain positive work history and vocational development	• Maintain positive work history and vocational development	• Assist patient with planning job search • Refer for career counseling • Reinforce strengths and resources	• Refer for career counseling • Develop strengths and resources	• Psychological testing to determine reasons for employment problems • Refer for vocational rehabilitation
Financial resources	• Reinforce responsible financial planning	• Reinforce responsible financial planning	• Review patient's budget • Bibliotherapy	• Refer for financial counseling • Family sessions to assess nature of problems	• Coordinate application for welfare, SSDI • Refer for financial counseling • Pursue guardianship for finances
Legal issues	• Reinforce responsible approach to legal and safety issues	• Reinforce responsible approach to legal and safety issues	• Individual therapy • Obtain copies of legal proceedings or reports	• Monitor attendance and performance at school or work • Monitor patient's appointments with court officials	• Enlist legal aid • Invite probation officials to periodic meetings • Refer patient to ex-offender programs
Military history	• If patient served in military, show gratitude for the service	• If patient served in military, show gratitude for the service	• Individual therapy	• Refer to VA	• Refer to VA • Refer to veterans support group
Activities of interest/hobbies	• Maintain activities and interests • Maintain healthy balance of work and leisure	• Encourage engagement in past interests • Explore new interests and hobbies	• Facilitate involvement in community activities • Explore new interests and hobbies	• Individual therapy regarding enjoyment and meaning in life	• Refer for occupational therapy • Meet with family or friends to coordinate activities
Religion	• Maintain meaningful religious involvements	• If interested, encourage engagement in past interests • If interested, explore new religious involvement	• Bibliotherapy • If interested, explore new religious involvement	• Consult with religious leader • Refer to religious leader or community	• Patient consult with religious leader • Bring religious leader into session to consult regarding issues

(Continued)

Table 9.3 (Continued)

Domains and Components	Strengths (+1 to +3)	No Problem (0)	Mild Severity (1)	Moderate Severity (2)	Severe (3)
Spirituality	• Maintain meaningful spiritual involvements	• If interested, encourage engagement in past interests • If interested, explore new spiritual involvement	• Bibliotherapy • If interested, explore new spiritual involvement	• Consult with spiritual leader • Refer to spiritual leader or community	• Patient consult with spiritual leader • Bring spiritual leader into session to consult regarding issues
Multicultural issues	• Maintain meaningful cultural involvements	• If interested, encourage engagement in past interests • If interested, explore new cultural involvements	• Bibliotherapy • Refer to community organizations	• Consult with multicultural leader • Refer to community organizations • Refer to multicultural leader or community specialist	• Patient consult with cultural leader • Bring cultural leader into session to consult regarding issues

examples of interventions that can be considered across the biopsychosocial domains. This listing is provided only to illustrate the wide range of interventions that are available for addressing strengths and weaknesses at varying levels of intensity. It certainly does not provide an exhaustive listing of interventions that can be integrated into treatment or any evaluation of the empirical evidence regarding the effectiveness of the interventions. (More complete discussions of these issues are available from Dziegielewski, 2010; Johnson, 2003; Jongsma, Berghuis, & Bruce, 2008; Jongsma, Peterson, & Bruce, 2006; Jongsma, Peterson, McInnis, & Bruce, 2006; Seligman & Reichenberg, 2007; and others).

Though treatment planning is a complex topic that requires extensive study and clinical experience to master, the factors discussed above provide the general framework for conceptualizing treatment planning from a comprehensive biopsychosocial perspective. Clearly, this approach is very different from traditional approaches that involve implementing the treatment (often the single therapy) that follows from one's adopted theoretical orientation. The present approach is far more complicated. This complexity is unavoidable, however, because patients' problems and circumstances are complicated. This complexity simply reflects the nature of human psychology.

The case example involving the 49-year-old male patient that was discussed in the last chapter is discussed here again in terms of treatment planning from a biopsychosocial approach. In the previous chapter, the biopsychosocial approach to assessment was contrasted with the cognitive-behavioral approach. Though some cognitive-behavioral therapists might employ a more flexible integrative approach that incorporates different techniques and methods into their treatment, strict cognitive-behavioral therapists are likely to use either Beck's cognitive restructuring approach (Beck, Rush, Shaw, & Emery, 1979) or Ellis' (1973) rational-emotive behavioral therapy approach. Both of these two approaches are well known, so there is no need to review the cognitive-behavioral treatment approach that would likely be used with this case example. The biopsychosocial approach to treatment planning is not as well known, however, so the case example presented in the previous chapter is elaborated on here in terms of the elements that would likely be considered as part of biopsychosocial treatment planning with this patient.

Case Example: A Biopsychosocial Approach to Treatment Planning With a Mildly Depressed Patient

One of the things the 49-year-old white male with mild depression said in his first session was that he was worried that he may not have any worthwhile reason to live after he retires. Though this did not suggest any suicidal ideation at the present time, and the patient denied having any, nonetheless the psychologist planned to follow up regarding this issue until she was satisfied that there were no signs of suicidal thoughts or feelings as treatment progressed.

The patient already stated that his feelings of emptiness about his work and family life may have derived from his upbringing and the example his mother and father set for him, so some exploration of his childhood and the modeling his parents provided was likely to be therapeutic. After the patient's wife joined the patient at the second session, it was also clear that continued discussions about their relationships with each other, their children, and their extended families would be productive.

The discussion with his wife also made it clear how the patient abandoned his former interest in maintaining his physical health. It was clear that this was a disappointment to her and also something that had a negative impact on the patient's self-image. Therefore, the patient agreed to improve his diet and initiate a physical exercise regimen. Because his wife was a physical therapist, he had ample information about how to go about this in a safe, healthy manner. The lack of meaning and importance the patient associated with his current work was another significant concern that emerged in the assessment, and so the psychologist wanted to explore those concerns further in individual sessions.

Therefore, the following elements were included in the treatment plan:

1. Monitor level of depression and any suicidal ideation that may emerge.
2. In conjoint marital sessions, discuss the evolution of the marriage and family life from when they started dating until the present, improvements that would be desirable with regard to his wife and children, and hopes and plans for the future.
3. In individual sessions, explore the nature of the patient's family-of-origin environment and build an awareness of the effect of his childhood on his adult approach to marriage, family, and work. Work toward a clearer differentiation of himself from his family of origin and clarify his approach to marriage, family, and work that is built around his personal values, beliefs, and goals.
4. In individual sessions, explore the satisfaction the patient derives from his current work and alternatives for improving the meaning and satisfaction he might derive from his work and other professional activities.
5. In individual sessions, explore family, leisure, cultural, or other activities that may provide meaning, engagement, and satisfaction in his life.
6. Resume a physical health program aimed at maintaining a healthy diet, increasing physical activity, and engaging in regular exercise.

10 Treatment

Behavioral health care from a biopsychosocial perspective is an involved, complicated process. As seen in the previous two chapters, the range of issues that falls
under psychological assessment and treatment planning is extensive, and treatment
from this approach consequently includes a wide variety of interventions and strategies as well. Not only do psychologists need to address patients' presenting problems, but they might intervene regarding multiple biopsychosocial problem areas
and build strengths in still other areas as part of a comprehensive, holistic approach
to addressing problems and promoting biopsychosocial functioning in general.
Broadening the focus of treatment to include functioning across the biopsychosocial domains means that treatment can be significantly more complicated than
some traditional approaches that focus on offering a specific type of therapy for a
proscribed set of issues, and largely leave other issues alone.

The purpose of the present chapter is not to examine all these issues, but to
examine a particular set of basic conceptual questions underlying the biopsychosocial approach to behavioral health treatment. A full discussion of empirically supported psychotherapy and other treatment approaches that can be incorporated into
behavioral health care from a biopsychosocial perspective would extend well
beyond a single chapter. To be competent at offering psychological treatments also
requires extensive bodies of knowledge and skill in other areas (e.g., knowledge of
normal and abnormal development, sociocultural factors, relationship- and alliance-
building skills, ethical and legal issues). Readers will need to rely on other
resources to obtain that information. Instead, the focus here is on the basic justification and rationale for providing behavioral health care from a biopsychosocial
perspective.

The main issues examined in this chapter follow from the health care emphasis
of the biopsychosocial approach to behavioral health care. Before providing behavioral health care to patients, one needs to be able to ensure the safety and effectiveness of the interventions that one might use with patients. This is required for
practicing in an evidence-based manner and also to meet the ethical obligations of
nonmaleficence and beneficence (i.e., to avoid harm and provide benefit). This
topic is widely discussed in professional psychology at a general level and was also
briefly reviewed in the previous chapter to help establish the basic justification for
treatment planning. It is reviewed in more detail here so that more specific

Foundations of Professional Psychology. DOI: 10.1016/B978-0-12-385079-9.00010-2

questions regarding the safety and effectiveness of psychotherapy can be examined. The specific questions addressed in this chapter include the following:

- Is psychotherapy effective?
  - Are the benefits provided by psychotherapy clinically significant? How often do patients return to normal functioning following treatment?
  - Do the benefits of psychotherapy last?
  - How does the effectiveness of psychotherapy compare with the effectiveness of psychiatric medications?
  - Does psychotherapy work for all patients?
- What factors account for the effectiveness of psychotherapy?
  - How important are the skills of the individual therapist in explaining therapy effectiveness?
  - Can the number of treatment failures be reduced?

The discussion of these questions will be limited to individual psychotherapy. Family and group therapy, combined medication and psychotherapy, bibliotherapy, computer-assisted interventions, support groups, and other interventions play important roles in behavioral health care, but other resources will need to be consulted to review the effectiveness of those treatment formats. The full range of behavioral health interventions is critical to the biopsychosocial conceptualization of psychological practice, but the present chapter will focus on the most common form of psychological treatment to keep the scope of the discussion manageable.

This chapter also discusses how treatment is different when approached from a biopsychosocial perspective compared with traditional approaches. A biopsychosocial approach tends to broaden the conceptualization of treatment compared to many traditional approaches to psychological practice. An overview of those differences is discussed later in the chapter.

Is Psychotherapy Effective?

In 1952, Hans Eysenck famously questioned the widespread assumption that psychotherapy was effective. It took a quarter-century for that question to be convincingly answered, and it is now widely considered to be settled. The data are consistent in showing that psychotherapy is generally effective for a broad range of mental health disorders and concerns and across a wide range of therapy approaches. The evidence supporting the effectiveness of psychotherapy is extensive. Smith and Glass (1977) conducted the first meta-analysis of the research regarding the effectiveness of psychotherapy and found an overall effect size of 0.85 (Smith, Glass, & Miller, 1980). Many meta-analyses followed, including enough to conduct meta-analyses of meta-analyses. Lipsey and Wilson (1993) reviewed all the meta-analyses available at that time and determined that the mean effect size for the controlled studies was 0.81. Lambert and Bergin (1994) conducted a similar analysis and found an average effect size of 0.82, while Grissom (1996) found an aggregate effect size of 0.75. Wampold (2001) examined the results from these and other meta-analyses and concluded that "A reasonable and defensible point estimate for the efficacy of psychotherapy would be .80... This effect would be classified as a large effect in the social sciences, which means that the average patient receiving therapy would be better off than 79% of untreated patients, that psychotherapy accounts for about 14% of the variance in outcomes, and the success rate would change from 31% for the control group to 69% for the treatment group. Simply stated, *psychotherapy is remarkably efficacious*" (italics in the original; pp. 70−71).

The effectiveness of psychotherapy is apparent when its effect sizes are compared to educational, psychopharmacological, medical, and other human service interventions (Barlow, 2004; Meyer et al., 2001; Reed & Eisman, 2006). In fact, the effect size of 0.80 for psychotherapy far exceeds that of many commonly accepted medical treatments (Meyer et al., 2001). For example, the correlation between coronary artery bypass surgery for stable heart disease and survival at 5 years is 0.08, between antibiotic treatment for acute middle ear pain in children and improvement at 2−7 days is 0.07, and between taking aspirin and reduced risk of death by heart attack is 0.02 (these would be categorized as small or exceedingly small effect sizes; Cohen (1988) concluded that $r = \pm 0.10$ is a small effect). When converted to another metric, the data suggest that 127 heart disease patients would need to be treated with aspirin before one death by heart attack is prevented (Wampold, 2007). This is in contrast to psychotherapy, where the majority of patients benefit from treatment. This does not suggest that treatments with very small effect sizes should be discounted. As long as the costs and risks of taking aspirin, for example, are minimal, the benefit to the very small number of individuals whose lives are saved by taking aspirin is great. In the case of psychotherapy, however, psychologists, other health care providers, and patients all can be assured that the overall effect sizes are large and a sizable majority of patients can be expected to improve following treatment.

Are the Benefits of Psychotherapy Clinically Significant?

Research indicates that psychological treatment provides a clinically meaningful improvement in patients' functioning and not just a statistically significant improvement. Indeed, large numbers of patients return to normal functioning following treatment. Most of the research examining this question relies on standardized measures of therapy outcome, and posttreatment scores falling to within one standard deviation of the normative mean suggest a return to normal functioning. Three meta-analyses have found that patients' average posttreatment scores on outcome measures moved into the range reflecting normal functioning (Abramowitz, 1996; Nietzel, Russell, Hemmings, & Gretter, 1987; Trull, Nietzel, & Main, 1988). After reviewing these and other studies, Lambert and Archer (2006) concluded that approximately three quarters of patients who undergo treatment show positive benefits, and 40−60% return to a state of normal functioning. In terms of patients in general, the benefits of psychotherapy are clinically quite meaningful. This question will also be addressed below in terms of evaluating the benefits of treatment in the individual case.

Do the Benefits of Psychotherapy Last?

The effectiveness of psychotherapy appears to be durable as well as clinically significant. Research examining the long-term effectiveness of psychotherapy is difficult because many patients drop out of follow-up studies or obtain other forms of
therapeutic intervention during the follow-up period. Nonetheless, studies have
tracked patients for up to 5 or more years following the end of treatment and have
consistently found that therapy improvements tend to endure (Lambert & Archer,
2006). There tends to be some decay in psychotherapy benefits over time for most
psychotherapies, though the decay is far less than for psychotropic medications, for
example (see the next section). In addition, the benefits of some therapies have
actually been found to increase over time. In five independent meta-analyses, effect
sizes for psychodynamic therapy at long-term follow-up were actually higher than
they were at posttreatment—the effect sizes at follow-up ranged from 0.94 to 1.57,
which are very large effect sizes (Shedler, 2010).

A recent long-term follow-up study that examined the effectiveness of therapy
for borderline personality disorder is notable because this disorder is commonly
considered to be one of the most difficult disorders to treat. This study also
included one of the longest follow-up periods of any study that has investigated the
maintenance of therapy gains, and the researchers were able to follow 100% of the
patients who originally entered the study. Bateman and Fonagy (2008) examined
therapy outcomes 5 years after patients finished a randomized, controlled trial comparing the effectiveness of psychodynamic therapy and treatment as usual. The psychodynamic group was found to have far lower rates of suicidality, further
outpatient treatment, and use of medication, higher DSM-IV-TR Global
Assessment of Functioning scores, and much improved vocational status. Only
13% of patients in the psychodynamic group still met the diagnostic criteria for
borderline personality disorder at the 5-year follow-up compared to 87% in the
treatment-as-usual group.

How Does the Effectiveness of Psychotherapy Compare with Medications?

Psychotherapy has been shown to be quite effective when compared to pharmacological interventions. While medications have often been considered the first line
of treatment for mental disorders in the medical community (e.g., Munoz, Hollon,
McGrath, Rehm, & VandenBos, 1994), psychological interventions have generally
been shown to be equal or greater in effectiveness than medicines for a range of
psychological disorders except for the most severe conditions such as schizophrenia
and bipolar affective disorder (Barlow, 2004; Elkin, 1994; Meyer et al., 2001;
Thase & Jindal, 2004).

Recent meta-analytic results are particularly informative for comparing the
effects of psychotherapy and antidepressant medication. An analysis of US Food
and Drug Administration (FDA) databases (including published and unpublished
studies) found that the overall mean effect size for antidepressants approved by the
FDA between 1987 and 2004 was 0.31 (Turner, Matthews, Linardatos, Tell, &
Rosenthal, 2008). The effect sizes ranged from 0.26 for fluoxetine (Prozac) to 0.31

for escitalopram (Lexapro). Methodological differences between medication and psychotherapy trials may be significant enough to prevent direct comparisons of the effect sizes between the two sets of studies, but the effect sizes for antidepressants are nonetheless relatively small. When the effects of antidepressants are compared to those of a placebo pill, antidepressants have also been found to be no more effective than placebo for mild, moderate, and severe depression (Fournier et al., 2010). Both placebo and antidepressant medication were associated with clinically significant improvements in depressive symptomatology scores in this meta-analysis. The effect of antidepressant medication was found to be superior to placebo only for those with very severe depression, a group that represents a relatively small number of those seeking treatment for depression.

Psychological interventions also have other important advantages over pharmacological approaches. Medicines frequently have significant unwelcome side effects, and surveys consistently find that the public prefers psychological to pharmacological interventions when they are given a choice (e.g., Hazlett-Stevens et al., 2002; Hofmann et al., 1998; Zoellner, Feeny, Cochran, & Pruitt, 2003).

A critical advantage of psychotherapy over pharmacological interventions, however, is the superior durability of the benefits of psychotherapy (Barlow, 2004). In the case of major depression, for example, medicines, placebo, and psychotherapy are typically helpful in reducing symptoms. Depressive episodes also tend to eventually remit on their own without treatment. The critical problem is that depressive episodes usually recur (Judd, 1997). Consequently, treatments need to prevent recurrence in order to be truly effective. Studies consistently find that psychological treatments typically provide durable benefits that last long after therapy is discontinued, while depressive symptoms often return once antidepressants are no longer taken (e.g., Hollon & Beck, 2004; Paykel et al., 1999; Teasdale et al., 2000). The same pattern of results has been found for anxiety disorders (Gould, Otto, & Pollack, 1995; Gould, Otto, Pollack, & Yap, 1997; Otto, Smits, & Reese, 2005). An important exception to this trend involves more serious biologically based disorders such as bipolar disorder and schizophrenia, where psychotherapeutic interventions are generally second in effectiveness to pharmacological ones (Lambert & Archer, 2006; Lambert & Ogles, 2004). Aside from these conditions, however, psychotherapy is the treatment of choice for many of the most common forms of psychological distress and disorder.

Does Psychotherapy Work for All Patients?

While psychotherapy is remarkably effective overall, there is substantial variability in the rate of improvement across patients. On one end of the continuum, a significant proportion of patients improve dramatically in just a short period of time. Though there has not been a large amount of research on this phenomenon, several studies have found that a significant proportion of patients make dramatic improvements in the first few weeks of treatment and this improvement is maintained at follow-up, which has been measured at up to 2 years posttreatment (Agras et al., 2000; Fennel & Teasdale, 1987; Haas, Hill, Lambert, & Morrell, 2002; Ilardi & Craighead, 1994; Renaud et al., 1998). This type of early dramatic response to treatment appears to be relatively common. Lambert (2007) estimated that perhaps 25% of patients are early responders and may not need treatment that extends beyond a few sessions. Not surprisingly, low severity of patient psychopathology is an important predictor of how quickly patients respond to treatment (Haas et al., 2002; Taylor & McLean, 1993).

Unfortunately, a number of patients do not appear to benefit from therapy. This is not unexpected, given the wide variability in the severity and complexity of patients' psychopathology. Some individuals are on a steadily declining trajectory of functioning that even the most effective therapies and therapists cannot reverse (the same is true in medicine, of course). Lambert and Archer (2006) have estimated that about 5−10% of patients actually deteriorate during treatment and an additional 15−25% do not measurably improve.

Even if patients are on a deteriorating course, slowing that deterioration would be a beneficial outcome of treatment. Nonetheless, patient deterioration is a major concern in the field, particularly given the evidence of potentially harmful psychological treatments (Barlow, 2010; Lilienfeld, 2007). Treatments such as rebirthing attachment therapy, conversion therapy for gay and lesbian individuals, critical incident stress debriefing, and grief therapy have been found to be potentially harmful (see Chapter 6). In addition, there is evidence that differences in therapist skill also affect therapy outcome (see the next section). Therefore, it is important that the outcomes of treatment are monitored so that patient deterioration is identified and appropriate adjustments are made (see Chapter 11).

What Factors Account for the Effectiveness of Psychotherapy?

What factors make psychotherapy effective have been debated throughout the history of the field. Early on, in fact, this was a major point of controversy in Freud's inner circle and resulted in the removal of Alfred Adler, who disagreed about the role of sexual instincts in personality functioning and the best approach to address neuroses (Gay, 1988). Heated controversies regarding the effective elements and processes for a wide range of psychotherapies have continued up to the present. Recently, however, better-controlled research designs are beginning to yield answers to some of these questions.

Psychotherapy clinicians and researchers have long hypothesized that several specific factors contribute to the effectiveness of psychotherapy. It appeared obvious that the skill and competence of the individual therapist was a significant factor. It has also long been believed that specific methods and techniques are more effective for certain disorders or personality characteristics than for others. Some psychologists have also long believed that there were factors common across therapies, such as therapist empathy, warmth, acceptance, and encouragement, that account for the effectiveness of therapy. It was also obvious that a major part of the reason that some patients improved while others did not was not related to the

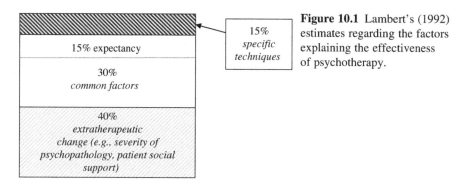

Figure 10.1 Lambert's (1992) estimates regarding the factors explaining the effectiveness of psychotherapy.

quality of the therapist or the treatment being offered, but rather had to do with characteristics of the patient (e.g., especially the severity of the psychopathology) or the patient's environment (e.g., positive and negative influences arising from the patient's family, support system, and community; see Garfield, 1994, for a thorough review of these factors).

After examining the available research on this issue, Lambert (1992) identified four elements that he believed accounted for the effectiveness of treatment (Figure 10.1). He estimated that the following proportions of the total variance in therapy outcome could be attributed to four factors:

1. Specific techniques (15%)—the effectiveness of particular treatments or techniques for treating particular disorders.
2. Expectancy (15%)—expectations that one will improve as the result of being in treatment; this has also been referred to as the placebo effect.
3. Common factors (30%)—factors found across therapies, such as empathy, warmth, acceptance, and encouragement to take risks.
4. Extratherapeutic change (40%)—factors associated with the patient (e.g., severity of psychopathology and level of ego strength) or the patient's environment (e.g., availability of social support).

Wampold (2001) analyzed the findings from several therapy outcome studies to obtain empirically based estimates of these factors. He concluded that very little of the variance in outcome is associated with specific ingredients associated with particular types of therapy. Instead, the competence of the individual therapist had the greatest effect on therapy outcome. Wampold estimated that only between 0% and 8% of the total variance in therapy outcome is accounted for by the specific type of treatment provided, while up to 70% of the variance in the effectiveness of therapy is attributable to the competence of the therapist. The rest of the variance in outcome was not explained by either of these two factors, though a significant portion likely consists of patient factors such as severity of psychopathology. Figure 10.2 depicts the proportions of therapy outcome variance accounted for by these factors.

Many sources of data suggest that therapist skill and competence are important to therapy outcome (see the next section). The question of whether specific treatments are more effective for treating particular disorders, however, is still

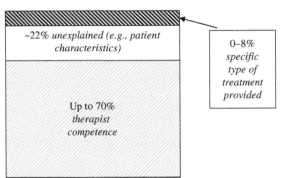

Figure 10.2 Wampold's estimates regarding the factors explaining the effectiveness of psychotherapy (adapted from Figure 9.2 in Wampold, 2001).

controversial. Wampold (2006) is among those who argue that there is no convincing evidence that particular therapies are superior to others as long as all the treatments being compared are *bona fide* therapies. Ollendick and King (2006), on the other hand, are among those who argue that some therapies are more effective than others or treatment as usual for treating particular disorders.

The field needs to wait for more research to resolve these questions. Research clearly suggests that therapist skill and competence are major factors accounting for the effectiveness of psychotherapy, as is patients' level of psychopathology when they enter treatment. In comparison, the type of therapy or theoretical orientation used with particular patients and disorders appears to have a much smaller impact on therapy outcome. It is still unclear, however, whether there are particular therapy techniques or foci, largely independent of theoretical orientation, which are important for explaining therapy outcomes (e.g., focusing on patients' emotional involvement and processing, regardless of the type of therapy employed, may be important to positive therapy outcomes; Shedler, 2010). The field will have to wait for more research to resolve this last question.

How Important Are the Skills of the Individual Therapist in Explaining Therapy Effectiveness?

Research clearly supports the conclusion that some therapist characteristics and behaviors contribute to positive therapy outcomes while others lead to poorer outcomes. Of course, this would be the expected finding—a relationship between practitioner competence and outcome undoubtedly exists in all professions. Nonetheless, voluminous research has confirmed that therapist qualities are indeed very important determinants of therapy outcome (for reviews, see Horvath & Bedi, 2002; Lambert & Barley, 2002; Norcross, 2002; Wampold, 2001).

The importance of individual therapist characteristics to treatment outcome was highlighted by the findings of two large overlapping studies that compared patient outcomes across 71 therapists who each treated a minimum of 30 patients (Okiishi et al., 2006; Okiiski, Lambert, Nielsen, & Ogles, 2003). The therapists were categorized in terms of the improvement seen in their patients' treatment outcomes.

Therapists who were in the middle 50% of the distribution tended to be largely indistinguishable from each other. At the extremes, however, distinct differences were found. The top 10% of the therapists who were associated with the best outcomes had an improved or recovered rate of 44% and a deterioration rate of 5%. The bottom 10% of therapists, on the other hand, had an improved or recovered rate of 28% and a deterioration rate of 11%. One particularly effective therapist who saw more than 300 patients had a deterioration rate of less than 1%, while a less effective therapist who saw more than 160 patients had a deterioration rate of 19%.

Additional research is needed to uncover the specific elements that account for these kinds of differences in therapy outcomes. It appears clear, however, that one of the critical elements is the ability to create therapeutic alliances and relationships. Even in pharmacotherapy provided by psychiatrists, the therapeutic alliance appears to be an important factor contributing to patient outcome. In an analysis of the NIMH Collaborative Depression Study, Krupnick et al. (1996) found that the quality of the therapeutic alliance was the most important factor explaining improvement in patients' depression in both the psychotherapy and pharmacotherapy conditions.

Norcross and Lambert (2006) summarized the research regarding the importance of the therapeutic relationship and concluded that "The therapy relationship makes substantial and consistent contributions to psychotherapy outcome for all types of treatments, including pharmacotherapy. Efforts to promulgate lists of evidence-based treatments without including the therapy relationship are thus seriously incomplete on both clinical and empirical grounds. Correspondingly, [evidence-based practices] should explicitly address therapist behaviors and qualities that promote a facilitative therapy relationship" (p. 217).

More research is needed to identify the specific therapist qualities that are associated with treatment outcomes (Beutler, et al., 2004; Wampold, 2006), but there is evidence for the importance of several factors. After summarizing the literature on this topic, Horvath and Bedi (2002) concluded that there are three categories of such therapist qualities: an interpersonal skill component, an intrapersonal component, and an interactive component. An interpersonal skills component relates to a therapist's ability to respond sensitively to patient needs, maintain open and clear communication, and communicate empathy and openness. It also includes avoiding taking too much control, offering interpretations prematurely, or being irritable. An intrapersonal component refers to negative and hostile therapist behaviors and dysfunctional personality characteristics, while the third component refers to the ability to develop a collaborative relationship with patients. The research reviewed by Shedler (2010) also suggests that the ability to get patients to effectively process emotion is important to therapy outcomes.

Can the Number of Treatment Failures Be Reduced?

While the field waits for more research on how to best develop the therapist qualities and characteristics that lead to effective treatment, a relatively simple approach has been identified that has potential for improving therapy outcomes across patients in general. Lambert and his colleagues have investigated the effect of providing therapists, and sometimes their patients, with patient outcome feedback regarding the ongoing progress of treatment. Using a standardized outcome measure to track patient symptomatology and level of functioning, patients were assigned (in four of the five studies, at random) either to the treatment-as-usual control condition or to the condition where their therapists received feedback regarding the patients' level-of-functioning scores (Lambert, 2007). For those patients who began deteriorating during treatment, the patients of therapists who did not receive feedback had posttest scores that were slightly worse, on average, than when they entered treatment. All of the groups where feedback was provided to the therapists, however, improved significantly by posttest. The effect sizes between those who received feedback versus the treatment-as-usual groups ranged from 0.34 to 0.92 across the five studies; these are surprisingly large effects given that the comparison group received actual psychotherapy.

Obtaining feedback regarding patients' progress during the ongoing course of treatment is not a new technique. Wolpe's (1958) systematic desensitization, for example, involves monitoring patient progress in treatment at a relatively molecular level. In this treatment, patients indicate the success of the counterconditioning on a moment-by-moment basis, often by raising their finger or verbally indicating their level of distress during the reciprocal inhibition sessions. Sobell and Sobell (2000) note that an inherent feature of the graduated nature of most substance abuse treatment, where more intensive treatment is provided as the severity of the problems increase, is that it is self-correcting. Patients' progress is monitored in an ongoing manner (e.g., urinalyses are frequently used to supplement the unreliability of patient self-report) and treatment strategies are adjusted to match deterioration or improvement in patients' progress.

Systematically monitoring patients' progress during the ongoing course of treatment has not been common in individual therapy, however. Therapy progress is routinely monitored by simply asking patients about how they are doing, and some behavioral and other treatments have fully incorporated the systematic measurement of treatment progress into treatment, but routine systematic outcomes assessment throughout behavioral health care has not yet occurred. Given the findings of research by Lambert (2007) and others, however, this issue is likely to receive increased attention in behavioral health care. One would expect that therapists and their supervisors would become more attentive when they receive information that their patients' symptoms and functioning are worsening and that they would then make adjustments in order to realize the best possible treatment outcomes. If research continues to support the effectiveness of this treatment strategy, the systematic monitoring of treatment progress may become a standard component of behavioral health care.

A Biopsychosocial Perspective on Treatment

The discussion above focused on the effectiveness of individual psychotherapy. Obviously, psychological treatment involves more than just individual psychotherapy, and the biopsychosocial approach in particular emphasizes a comprehensive, integrative approach that extends well beyond individual psychotherapy. The broadening of case conceptualization and the treatment options that can be incorporated into one's clinical practice are among the important differences between a biopsychosocial approach and many traditional approaches to behavioral health treatment. Incorporating a broader range of biopsychosocial interventions into treatment typically also requires that psychologists are able to work collaboratively with multidisciplinary treatment teams. The emphasis on monitoring the progress of treatment to help ensure that psychologists are meeting the biopsychosocial needs of their patients is another important result of applying the health care orientation of the biopsychosocial approach.

Broadening of Case Conceptualization

Historically, graduate education in professional psychology often emphasized the learning of psychodynamic, behavioral, cognitive, and humanistic approaches to individual psychotherapy, and the focus of treatment was often limited, in many cases emphasizing depression and anxiety. Certainly many additional treatment interventions have been taught across programs, but the curriculum in many programs tended to emphasize these areas. In contrast, a biopsychosocial approach to case conceptualization emphasizes a holistic, systemic, and developmental perspective to understanding patients, addressing their behavioral health needs, and improving their biopsychosocial functioning in general.

The biopsychosocial perspective focuses attention on the full range of behavioral health and other biopsychosocial concerns that are important in patients' lives. Chapter 3 summarized the range of biopsychosocial problems and concerns that are common in the general population. For example, common medical conditions such as obesity, arthritis and other pain conditions, insomnia, pregnancy, and many others (see Table 3.2) clearly have substantial impacts on individual and family functioning, and psychological and behavioral factors are involved in the etiology, consequences, and/or treatment of most of these conditions. In the psychological domain, many patients struggle with addictions, a history of child abuse, personality dysfunction, and several additional topics that are not always heavily emphasized in professional psychology training. Many issues in the sociocultural domain are also very important in patients' lives but are addressed relatively little in graduate training, such as relationship skills, parenting for those with children, financial stress and vocational instability, racism and sexism, criminal involvement and victimization, single parenthood, and divorce and reconfigured families. The co-occurrence of problems within and across these domains is also very common and is often not emphasized extensively in graduate training.

The biopsychosocial approach also incorporates a health, strengths, and wellness perspective, and optimizing functioning across the biopsychosocial domains is a priority from this perspective. Even when patients have no problems or have mild strengths in particular areas, converting these to major strengths obviously can be beneficial.

The biopsychosocial approach also emphasizes a long-term developmental perspective in addition to focusing on current functioning. A developmental perspective is necessary for fully understanding patients' personality characteristics and psychopathology, the etiology of their problems, and the role of risk, compensatory, and protective factors on their development and functioning. Monitoring and providing care for ongoing problems and conditions as well as building strengths and improving resilience over the long term are all priorities from this perspective (e.g., through building strong physical health, coping resources and resilience, social support, family functioning, educational and vocational effectiveness, financial stability). While biopsychosocially oriented psychologists commonly provide short-term, time-limited care for acute conditions and circumscribed problems, they also apply a long-term developmental perspective to comprehensively conceptualize patient cases.

Broadening of Treatment Options

From the biopsychosocial metatheoretical perspective, the current research literature suggests that there are a variety of therapeutic interventions available to address patient needs and maximize their biopsychosocial functioning. Though it is not possible to develop expertise with a large number of these treatment options, psychologists do need to have some familiarity with a range of interventions in order to work effectively from the biopsychosocial approach.

Practice guidelines do not yet exist to indicate the range of treatments that should be integrated into several types of general and specialized practice, but two trends now under way may become more prevalent in psychology practice. Psychologists are increasing the number of treatments they provide themselves, and this trend is likely to continue—psychologists' endorsement of eclectic and integrative approaches to practice has grown steadily over recent decades (Norcross, 2005). Small numbers of psychologists have even begun offering psychopharmacological services in addition to psychotherapeutic interventions (e.g., in Louisiana, New Mexico, and the US Department of Defense; Fox et al., 2009). It is also likely that graduated treatment approaches will increasingly become integrated into psychologists' repertoire of skills. These range from minimal to intensive treatment options to match the severity and complexity of patients' mental health concerns and needs (e.g., SBIRT and PLISSIT, see above). Already the United Kingdom has begun implementing a countrywide effort to increase access to mental health care by providing the level of mental health intervention that is efficient and effective for meeting the public's mental health needs across levels of severity and complexity (Clark et al., 2009).

There are also several types of practice settings where providing psychotherapy to address mental health issues is often not the first priority. In correctional, educational, and employment settings, for example, individuals' abilities to function and meet the goals and needs of the institution or organization they are a part of may be the first priority. Medical stabilization and treatment adherence are often top priorities in medical psychology as well. The biopsychosocial approach can easily accommodate this broad range of objectives and goals and their associated treatment interventions.

Increased Collaboration with Other Professionals and Third Parties

Individual psychologists obviously cannot be competent at treating all the biopsychosocial problems that patients have. As a result, they need to be able to work collaboratively with other professionals to address those needs. For example, the graduated treatment models just mentioned above tend to include brief, limited interventions for less serious problems and concerns, and these can be relatively easily learned and implemented. Interventions for more serious and complex problems require more extensive expertise, however, and so referral to and collaboration with specialists is often needed at these higher levels of treatment intensity.

Psychologists working with the various institutions referred to above (i.e., in correctional, educational, employment, and medical settings) often focus on supporting the priorities of those institutions as opposed to focusing solely on the treatment of mental disorders, as is common in behavioral health clinics. This requires that they work collaboratively with the other professionals in those institutions to meet the institutional goals as well as their patients' personal needs and goals.

Therapists working from the biopsychosocial perspective also tend to increase the involvement of family members, educators, employers, and others who may be important in the patient's life and who can contribute to positive treatment outcomes. Chapter 8 emphasize the importance of relying on a variety of collateral contacts such as family members to obtain reliable, useful assessment information. In many cases, these individuals can be integrated effectively into treatment as well. This is especially important as problem severity and complexity increases (e.g., serious substance abuse, psychiatric disorders, medical conditions, relationship and family problems) and when patients themselves are more vulnerable and dependent (e.g., children, individuals with cognitive disability). As level of care increases, the number of different types of professionals and third parties involved tends to increase, and the level of collaboration needed to effectively provide the care increases as well. Collaborative approaches are also advocated as necessary for effectively treating the large number of behavioral health needs found among infants, children, and adolescents in the United States (Egger & Emde, 2011; Kazak et al., 2010). This approach is also consistent with the recent promotion of integrated health care models where multidisciplinary collaborative care teams integrate behavioral health, disease management, and prevention services into primary care settings (American Psychological Association Presidential Task Force on the Future of Psychology Practice, 2009).

Systematic Monitoring of Treatment Outcomes

The health care emphasis of the biopsychosocial approach increases the attention focused on meeting the behavioral health and biopsychosocial needs of patients, which in turn increases the focus on the effectiveness of treatment. Treatment effectiveness cannot be properly assessed without systematically measuring treatment outcomes, so a biopsychosocial approach to professional psychology places more attention on outcomes assessment than most traditional approaches to behavioral health care. This is important during the course of treatment (e.g., to detect cases involving deterioration or no improvement) as well as at the termination of treatment. Chapter 11 provides an overview of the important issues involved in that topic.

Conclusions

The effectiveness of psychotherapy is impressive. The effect sizes of psychotherapy are large, particularly in relation to those for many medical, educational, and other human service interventions. The effectiveness of psychotherapy is also very meaningful clinically, enabling large numbers of individuals to return to states of normal functioning. These effects also tend to be enduring. Indeed, evidence suggests that not only do the benefits of therapy normally continue beyond the end of treatment (unlike the benefits of many psychotropic medications), but that the positive effects of some psychotherapies may even increase over time. A significant number of patients improve very quickly in therapy (e.g., in just a few sessions), though not everyone improves and there is also a small proportion who deteriorate during treatment. There is evidence, however, that the ongoing monitoring of treatment outcomes can turn around some of these cases.

Research on the mechanisms of therapy that account for treatment effectiveness is still in its early stages. Nonetheless, research clearly points to the importance of therapist skill in creating therapeutic relationships and alliances as critical to the overall effectiveness of psychotherapy. Wampold (2001) has estimated that as much as 70% of the variance in therapy outcomes is attributable to this factor. Certainly another important factor is the patient's prior level of psychopathology— the severity and complexity of patients' problems has a significant impact on whether treatment is effective. There is some initial evidence, however, that even complex, serious disorders can be treated effectively with psychotherapy. The question of whether certain therapies are more effective in treating certain disorders than other therapies is still controversial.

A biopsychosocial approach to behavioral health care emphasizes a broad, holistic, systemic, and developmental perspective to understanding treatment. A health and wellness perspective complements the traditional focus on problems and disorders, while the developmental perspective emphasizes etiology, the development of personality characteristics and psychopathology as well as strengths and weaknesses, and the role of risk, protective, and compensatory factors in treatment and behavior change.

Given the broad metatheoretical perspective of the biopsychosocial approach, a wide range of treatment options is also incorporated into this approach. Though it is impossible for psychologists to be competent with all the interventions used to treat the issues that patients deal with, a practical approach for increasing the number of treatments that could be applied involves the use of graduated or "stepped" models of treatment where less intensive interventions are recommended for less serious problems and concerns and more intensive interventions and referrals to specialists are recommended for more serious problems. Of course, the specialization in which one practices greatly affects the range and types of treatments one learns and uses. Research is still ongoing, however, regarding the most effective techniques and therapies for different populations and disorders. Psychologists need to monitor the research literature for guidance on the evidence-based treatments that can be safely and effectively used in different types of practice.

Across all types of settings within which psychologists work, practicing from a biopsychosocial approach is likely to increase collaboration with other types of human service professionals and others who are important in patients' lives such as family members, employers, educators, and medical providers. Fortunately, using the same biopsychosocial framework as other health care and social service professionals facilitates communication and collaboration with those individuals.

The case example that was discussed in the previous two chapters reappears below to illustrate the biopsychosocial approach to behavioral health care treatment. The treatment plan discussed in the previous chapter was implemented with the 49-year-old married male with mild depression. That plan included six components: (1) monitor the depressed mood and any emergence of suicidal ideation; (2) begin conjoint sessions with the man and his wife; (3) conduct individual sessions with the man alone to explore the influence of his own family-of-origin history, etc.; (4) explore his vocational satisfaction through individual sessions; (5) explore family, leisure, cultural, and other activities that may provide meaning, engagement, and satisfaction; and (6) improve his physical health through a healthier diet and exercise.

Case Example: A Biopsychosocial Approach to Psychotherapy with a Mildly Depressed Patient

The psychologist monitored any suicidal ideation that might emerge because of the patient's statement in his intake session that he did not know what reasons he would have to live for after retirement. The patient often acknowledged that he wondered if he would commit suicide if he felt that way at that time, but he denied having any suicidal thoughts or feelings at all in the past or at present.

The patient and his wife had very productive conversations when she came in to his second, third, and sixth sessions. When she came in the first time, the distance between them was evident—they interacted in a very polite and friendly manner, but they displayed little affection or closeness. He admitted that he did not find child-rearing duties rewarding, even though he loved his children very much, and that it is probably why he gradually spent more and more time following sports. She said that she wanted to continue doing the other things they used to do, such as going to see theater, film, music, and art. Since he frequently declined her suggestions to go to these events, she began going with her girlfriends instead. This has been enjoyable for her, but she knows that they have grown apart in terms of enjoying the time they spend together. The psychologist asked both members of the couple if either had thoughts of separation or divorce, and both maintained they have not entertained those kinds of thoughts.

The mood was much lighter in the couple's second session. They joked more frequently and their nonverbal behavior was less defensive. They enjoyed reminiscing about how much fun they had early in their relationship. She emphasized that she was very impressed that he had started exercising and improving his diet. "He actually threw out all the chips and cookies in the house. He ate up all the beef jerky, but at least he didn't buy any more. I've tried to get him to do this for years! Finally he did it—I can hardly believe it." The couple agreed to start going out on dates again, and several sessions later the husband noted that he was spending much less time watching sports on TV, but the other things he was now doing were actually more rewarding.

Before the fifth session, the patient retook the same questionnaire that he completed at the initial session. His scores this time suggested that his depressed mood had improved significantly and he was more satisfied with his level of functioning in different areas.

The patient spent several sessions exploring his memories of his childhood, his parents' own relationship, and their relationships with their families of origin. He was initially angry that his parents had not been warmer and more caring with him, more interested and attentive in his school and other activities, and that they did not talk with him about his future career plans and family hopes. He reported that he was also angry with himself because it was clear to him that he had fallen into a very similar pattern as his own father in terms of both work and family life.

By the seventh session, the patient began developing a different perspective on his parents. Both had grown up in relatively adverse conditions in poor Depression-era families that used fairly harsh approaches to child rearing. The patient decided to talk with his parents about their own childhoods, and he learned that both experienced what would now be considered at least mildly abusive upbringings—their parents commonly relied on physical discipline and shaming to control their children's behavior. After these conversations, the patient grew confused because he still felt anger at the emotionally neglectful and confusing nature of his own upbringing, but at the same time he was beginning to feel admiration for his parents' ability to provide a far healthier and caring family environment than what they themselves had experienced. After two more sessions, his admiration grew for what they had overcome and achieved in their own lives, and he said he did not feel it was reasonable to hold them accountable for their lack of knowledge about contemporary approaches to parenting—they actually did quite well given the ways they had learned to raise children and care for a family.

After 12 weekly sessions (either individual or conjoint with his wife), the psychologist recommended that they drop to every other week in frequency since he had made good progress working through many of the relevant issues. He still wanted to explore his work and career situation, and he was growing more concerned about the relationships that he had developed with his own children. These topics were the focus of most of the next four bimonthly sessions.

11 Outcomes Assessment

Professional psychologists' feelings about conducting outcomes assessment are often mixed. On the one hand, psychologists have welcomed the research findings showing that their treatments are effective—indeed, frequently more effective than many medical, educational, or other human service interventions. Psychologists are also quite comfortable with the technology of outcomes assessment which involves relatively simple psychological testing and provides relevant and useful information.

On the other hand, outcomes assessment has not been enthusiastically adopted by many practicing psychologists. Psychotherapy is a very private process and many psychotherapists are concerned about the safeguards in place to prevent the misuse of outcome data by managers or insurers. Therapists are sometimes also concerned about the additional workload for patients if they need to complete additional assessment instruments during and at the end of treatment. It is also natural for people to be uncomfortable with having their work evaluated, even though such evaluations have become common in many areas (e.g., student evaluations of college teaching, customer satisfaction surveys for many products and services). Similar types of evaluation have not been common in psychotherapy, however.

The movement to incorporate outcomes evaluation into health care and other human services is nonetheless moving ahead steadily (Maruish, 2004b; Ogles, Lambert, & Fields, 2002). Society increasingly demands accountability in general, and health care organizations are under growing pressure to ensure that services are effective and efficient. In addition, few would disagree with the basic reasons for incorporating outcomes evaluation into behavioral health practice, which include the need to ensure quality care, improve clinical practice, strengthen clinical science, and maintain the general ethical commitment to quality services (Barlow, Hayes, & Nelson, 1984; Bloom & Fischer, 1982; Maruish, 2004b).

Growing Importance of Outcomes Assessment

Outcomes assessment has only recently become integrated into mainstream behavioral health practice, even though it has a long history in the field. Fenichel was one of the first to study therapy outcome nearly a century ago (Bergin, 1971). Psychoanalysts at the Berlin Psychoanalytic Institute were asked to rate the outcome of their patients seen between 1920 and 1930. The patients themselves

Foundations of Professional Psychology. DOI: 10.1016/B978-0-12-385079-9.00011-4

provided no information regarding how much they improved from treatment, but Fenichel reported that the ratings provided by the analysts indicated that 91% of the patients could be classified as improved.

Most of the research on therapy outcome in subsequent decades focused on between-group studies designed to examine whether various psychotherapies were effective in generating improved outcomes compared with control groups. The meta-analytic technique was developed to aggregate the data from these individual studies to allow a comparison between treatment and control groups across studies and evaluate whether a whole body of research data indicated that particular treatments were effective or not. Relatively little attention was given to the question of whether the effectiveness of psychotherapy should be assessed on an individual patient basis, however, until the 1990s. In 1996, Ken Howard, a leader in this area, and his colleagues captured the changing mindset about conducting outcomes assessment on a routine basis with patients when they noted that "it is not sufficient for the practitioner to know that a particular treatment can work (efficacy) or does work (effectiveness) on average ... The practitioner needs to know what treatment is likely to work for a particular individual, and then whether the selected treatment is working for this patient" (p. 1060).

The techniques used for therapy outcome assessment evolved significantly over the years. The earliest studies (like the one by Fenichel) tended to use global ratings of patient improvement that were made by therapists at the termination of therapy and usually included no long-term follow-up (Lambert, Ogles, & Masters, 1992). A wide range of instruments were also used to measure outcome; many of the instruments were not standardized or included only one item (Froyd, Lambert, & Froyd, 1996; Meier & Davis, 1990). Further, perspectives on the effectiveness of treatment were found to vary depending on the source of the outcome data (e.g., obtained from the patient, therapist, significant other), and these data were often found to be uncorrelated with one another (Miller & Berman, 1983). These problems remained largely unresolved until the 1990s.

Interest in behavioral health outcomes assessment grew rapidly in the 1990s when research on reliable and efficient ways to gather useful outcome data progressed quickly. Ogles et al. (2002) declared that "Certainly the 1990s will be referred to in behavioral health service history as the decade of outcomes" (p. 1). Since then, the movement toward evidence-based practice has received greater attention than outcomes assessment specifically, but both are part of the same movement toward accountability and quantifiable evidence regarding the effectiveness of behavioral health services—evidence-based practice is primarily concerned with the effectiveness of treatment for groups, while outcomes assessment is focused on the effectiveness of treatment for the individual. The need for accountability and evidence-based practice is also consistent with the biopsychosocial conceptualization of professional psychology as a health care specialization that meets the behavioral health needs of the general population. If the field is conceptualized as primarily a service industry that offers a variety of services selected by individual patients to meet their own personal needs, then there is much less need to systematically measure outcomes, and accountability is managed primarily by the market. In health care, however, demands for systematic outcomes assessment and accountability are generally higher.

One focus of concern in the recent emphasis on accountability in health care involves safety and risks of harm to patients that are caused by health care services. The influential 2001 Institute of Medicine report entitled *To Err Is Human* estimated that 44,000−98,000 deaths occur annually in the United States as a result of medical errors. While most of these deaths are due to medical errors involving medications, misdiagnosis, and infections acquired while receiving medical care, the failure to accurately diagnose depression and suicide risk is another health care error that may contribute to a significant number of these deaths (Schiff et al., 2009). There was also widespread concern in the 1990s regarding the potential harm caused by therapy involving repressed memories of child abuse (Loftus & Davis, 2006). Several additional psychological treatments have also been identified in recent years as carrying the potential to cause harm in patients (Barlow, 2010; Lilienfeld, 2007; see Chapter 6).

Another major concern in American health care recently has been its cost, particularly when considered in relation to the outcomes associated with that care. More money is spent on health care in the United States per capita than in any other nation in the world, and a greater percentage of gross domestic product is spent on health care in the United States than in any other nation except for East Timor (World Health Organization, 2009). Despite having the most expensive health care in the world, outcomes involving population morbidity and mortality are low. Indicators of long, healthy, and productive lives are lower than in many other developed countries (Davis, Schoen, & Stremikis, 2010). In fact, the *World Health Report 2000* found that the United States ranked 37th in overall performance (France, Italy, Spain, Oman, Austria, and Japan were the top sizable countries; World Health Organization, 2000). The managed care revolution in the United States was supposed to bring efficiency in terms of implementing accountability and productivity models employed in industry, but that effort has not been particularly successful.

Another perspective on accountability in health care emphasizes professionals' ethical obligations to evaluate the effectiveness of their clinical services (Bloom, Fischer, & Orme, 2003). A laissez-faire or market approach to psychological services tends to emphasize that consumers are responsible for their decisions to purchase services and gives less responsibility to the service provider for ensuring that the services provided are appropriate and effective. The biopsychosocial approach advocated in this volume, however, emphasizes the ethical obligations of nonmaleficence and beneficence. Establishing the safety and effectiveness of one's services through systematic outcomes assessment helps ensure that one has met the obligations to avoid harm and provide benefit.

The APA Ethics Code does not include a specific responsibility to evaluate psychological services for effectiveness. It does, however, include the requirement that psychologists terminate therapy when it becomes reasonably clear that the patient no longer needs the service, is not benefiting, or is being harmed (see Standard 10.10, APA, 2002). As was noted in the previous chapter, Lambert and

Archer (2006) estimated that 5−10% of patients deteriorate during treatment and an additional 15−25% do not measurably improve. Relying on patient self-report alone to identify these cases may result in less reliable information than what is needed (e.g., if patients do not report negative feedback because they do not want to embarrass or disappoint their therapists). It is noteworthy that expectations to evaluate one's practice are clearer in some other behavioral health specializations. For example, the Code of Ethics of the National Association of Social Workers states that "Social workers monitor and evaluate ... practice interventions" (Standard 5.02(a), National Association of Social Workers, 2008). The American Counseling Association Ethics Code (2005) states that "Counselors continually monitor their effectiveness as professionals" (Code C.2.d).

Growing expectations for accountability, tighter health care budgets, and increased attention to the safety and effectiveness of treatments will increase pressures to monitor outcomes in behavioral health care. The health care orientation of the biopsychosocial perspective also emphasizes an evidence-based approach to monitoring treatment effectiveness. Fortunately, the technology of outcomes assessment has developed to the point where it can be efficiently integrated into professional psychology practice and can provide reliable and useful information regarding patients' progress in treatment.

This chapter provides a brief overview of outcomes assessment in health care generally before outlining the implementation of outcomes assessment in behavioral health care. Following that is a discussion of a biopsychosocial perspective on outcomes assessment. Taking these various factors into account then results in a summary of best practices in this area. Because this topic is much narrower than the subject of treatment covered in the previous chapter, more practical detail is included here in contrast to the foundational issues addressed in the previous chapter. This detail will be informative to many readers because professional psychology education and practice is only beginning to thoroughly incorporate outcomes assessment into the treatment process.

Outcomes Assessment in Health Care Generally

At the most basic level, treatment outcome simply refers to the result or consequence of a treatment. Outcomes assessment is valuable because it generally focuses on what is most meaningful to patients and other stakeholders. Outcome measures typically assess factors such as lessened symptomatology, improvements in functioning, improved quality of life, satisfaction with services delivered, and cost-effectiveness.

There are a variety of different approaches to assessing outcomes in health care. Medical researchers differentiate between two general classes of outcomes (Kane, 2006). Condition-specific measures typically focus on the symptomatology that reflects the status of the medical condition a patient has or the consequences that a disease has on a person's life. Generic measures, on the other hand, provide comprehensive assessments of health-related functioning across domains in a person's life that are not specific to a particular disease or condition (Maciejewski, 2006).

Generic Measures

Generic measures are designed to assess a full range of physical, psychological, and social aspects of health that were identified in the broad World Health Organization's (1948) definition of health. This definition focused on the *quality* of health in addition to its *quantity* in terms of life span and other easily quantified indicators. Measures emphasizing quantity of health generally focus on morbidity, mortality, and life expectancy, whereas measures emphasizing quality of health focus on overall health and functioning. Generic outcome measures (usually called global measures in professional psychology) tend to focus on patients' perceptions of their physical, psychological, and social functioning and their overall quality of life, factors that are often more relevant to patients than their condition-specific outcomes (Maciejewski, 2006).

Quality of life would appear to be critical to any measure of life satisfaction or the overall outcome of health care, but it can be difficult to define because individuals place very different priority on different aspects of their lives. Therefore, measures of quality of life typically include several domains of functioning. Eight dimensions are often considered critical to a comprehensive assessment of quality of general life functioning: physical functioning, social functioning, emotional functioning, sexual functioning, cognitive functioning, pain/discomfort, vitality, and overall well-being (Maciejewski, 2006; Patrick & Deyo, 1989). The 36-item Short-Form Health Survey (also known as the SF-36; Ware & Sherbourne, 1992) measures all eight of these dimensions and has become the most widely used generic measure of functioning in medical research (Maciejewski, 2006).

Condition-Specific Measures

Condition-specific measures of health care outcome are designed to assess aspects of functioning that are closely related to a patient's disease or condition. Condition-specific measures are designed to be highly sensitive so that they can detect even small treatment effects. There are two basic types of condition-specific outcome measures: clinical measures focus on the signs, symptoms, or test results associated with a particular disease or condition, and experiential measures focus on the impact of the disease or condition on the patient (Atherly, 2006). Generic health outcome measures can easily miss clinically important changes associated with treating a specific condition, whereas condition-specific measures are designed to provide clinically meaningful measures of treatment responsiveness, even in the absence of changes in overall functioning. For example, a successful treatment for hypertension is often imperceptible to patients because changes in blood pressure are difficult to detect. A generic outcome measure will likely fail to detect the effect of a successful or a failed treatment, whereas a blood pressure measurement will. Even when patients cannot detect the effect, a successful treat-ment for hypertension can have a profound influence on the long-term health of the patient (Atherly, 2006). Therefore, both generic and condition-specific measures are needed to thoroughly evaluate medical outcomes.

Outcomes Assessment in Behavioral Health Care

Outcomes assessment in behavioral health care often relies on a similar combina-tion of condition-specific and generic measures for the same types of reasons that medicine does. The outcomes of health care treatment, whether medical or psycho-logical, need to be judged in a multifaceted manner because individuals' lives and functioning are complex, particularly when viewed from a biopsychosocial perspec-tive. Focusing on single dimensions of functioning (e.g., on only decreasing the symptoms of a disorder) can simplify an assessment so that its meaningfulness is limited. Maruish (2004b) noted that "outcomes" is commonly used in its plural form to emphasize the importance of taking a multifaceted approach to outcomes assessment where the treatment effects are measured across multiple domains of functioning.

Selecting Outcome Measures

Ogles, Lambert, and Masters (1996) presented a comprehensive categorization of the characteristics of outcome measures that can be used to help guide the selection of instruments for evaluating a behavioral health program or one's clinical practice. They recommend that the following five categories be considered in this process:

- *Content*—typically focuses on cognitions, affect, and/or behavior; most instruments cover all three of these types of contents
- *Social*—change that occurs at the interpersonal versus the intrapersonal level, or at the level involved in performing one's social roles (e.g., at work or within the family)
- *Source*—the source of information (e.g., whether it is provided by the patient, therapist, family member, or an institutional official)
- *Technology or methodology*—methods used to gather the data; some instruments focus on global change, others on targeted behaviors or symptoms, others on observed behaviors, and others on measures of status (e.g., hospital discharge, recidivism)
- *Timeframe*—when the assessment is conducted, varying from pretreatment and posttreat-ment, at every session, periodically, or only at follow-up

Selecting outcome measures becomes a daunting task when considering all the possibilities for conducting outcomes assessment across the five dimensions identi-fied by Ogles et al. (1996). In addition, patient satisfaction with services and the cost-effectiveness of services are important areas for assessment as well. Some basic considerations help narrow the possibilities quickly, however.

The following three considerations are important when selecting psychological measurement instruments for any purpose, but particularly with regard to outcomes assessment:

1. **Outcome measures need to provide information that is useful for the purpose.** The selection of outcome measures varies across therapists and patients. The instruments selected for monitoring the effectiveness of the treatment for attention-deficit/hyperactivity disorder, for example, may have no overlap with instruments selected for evaluating treatment for sexual concerns, substance dependence, or severe mental illness.
2. **Be practical in terms of the time and monetary demands involved.** Outcome measurement needs to be practical from the perspective of both patients and therapists.
3. **Maximize the usefulness and interpretability of the information obtained.** It is highly recommended that clinicians use standardized instruments for assessing therapy outcome. The reliability and validity of major standardized instruments have been established through numerous research studies, and normative comparison data are typically also available. These are both valuable aids for interpreting the data obtained.

These three guidelines emphasize the practicality and usefulness of outcomes assessment. An additional important consideration with regard to the selection of instruments concerns whether it is most useful to select a global, condition-specific, or individualized measure of treatment outcome. Each has distinct advantages.

Global Measures

The most practical outcomes measure for many therapists and agencies will be a global or generic measure. Such measures can often be used with all the patients receiving services within a practice, and they also provide meaningful information regarding patients' general level of distress and overall functioning. There are several brief, standardized measures that serve these purposes well (Ogles et al., 2002). Widely used and respected instruments include (1) the Brief Symptom Inventory (BSI; Derogatis & Melisaratos, 1983); (2) the Outcome Questionnaire-45 (OQ-45; Lambert et al., 1996); and (3) the Short Form-36 (SF-36; Ware & Sherbourne, 1992). The four-item Outcome Assessment Scale was recently developed as a very brief alternative that is very easy to use in clinical practice (Bringhurst, Watson, Miller, & Duncan, 2006). A widely used instrument for children and youth is the Child Behavior Checklist (CBC; Achenbach & Edelbrock, 1983). The BSI focuses primarily on psychological symptoms, while the SF-36 and the CBC include scales measuring variables from across the biopsychosocial domains. Of course, there are many alternatives in addition to these instruments.

Condition-Specific Measures

Global or generic outcome measures often need to be supplemented by condition-specific measures to adequately assess the effectiveness of treatment for a particular disorder or problem. Whereas generic assessments provide useful information for assessing changes regarding general areas of functioning, more specific measurement is needed to evaluate changes regarding particular symptoms or behaviors. In fact, obtaining specific assessment information is critical to all phases of the treatment process. For example, at intake, using condition-specific measures (1) helps ensure that the psychologist understands what the patient is dealing with and that the patient knows that the psychologist understands; (2) provides information necessary for completing the assessment and making proper diagnoses; (3) provides information necessary for treatment planning; and (4) provides baseline data, which are necessary for evaluating the progress and the overall effectiveness of treatment. The specific signs and symptoms of individual disorders or problems (e.g., substance abuse, obsessive–compulsive disorder, enuresis, psychotic disorders, eating disorders, child abuse) are often missed entirely by generic intake and outcome measures.

There are many standardized instruments with strong psychometric properties that can be used for intake and outcomes assessment for specific disorders and problems. Psychologists working in specialized areas will become familiar with the instruments commonly used in their respective areas. Examples of instruments widely used in general practice for monitoring specific symptoms include the Beck Depression Inventory (BDI-II; Beck, Steer, & Brown, 1996), the State-Trait Anxiety Inventory (Spielberger, Gorsuch, & Lushene, 1970), the Fear Questionnaire (Marks & Mathews, 1978), and the Dyadic Adjustment Scale (for measuring relationship satisfaction in couples; Spanier, 1976). A wide range of additional instruments can be considered for this purpose depending on the specific issues and disorders involved.

Individualized Measures

In addition to using standardized instruments, it is frequently important to use individually tailored measures designed to assess the unique experiences or concerns of individual patients (Clement, 1999; Ogles et al., 2002). The uniqueness of each patient's concerns, circumstances, and biopsychosocial functioning needs to be evaluated during intake assessment, and an outcomes assessment focused on those same issues needs to be completed at the end of treatment to measure the effectiveness of treatment for that individual. One individual with substance dependence, depression, or agoraphobia may share few similarities with another person with the same disorder. Intake and outcomes assessments often need to be individualized as a result.

There are several approaches to conducting individualized assessment. One well-known approach involves the use of target complaints, a system included in the National Institute of Behavioral Health Core Battery Initiative (Waskow & Parloff, 1975). In this system, the patient, therapist, or the patient and therapist together identify targets for treatment and then rate the level of problem severity for each target complaint. A unique list of complaints, which might even seem to be unrelated to each other (e.g., social isolation, missed work days, shoplifting, arguments with one's mother), can then be tracked periodically starting from intake. Goal Attainment Scaling (Kiresuk & Sherman, 1968) uses a similar approach: treatment goals are established collaboratively by the patient and therapist, and progress toward meeting those goals is then assessed across treatment. Behavior therapy has also developed sophisticated approaches to measuring behaviors and assessing the effectiveness of treatment. Identifying target behaviors and performing a functional analysis can be a highly useful approach to intervention and outcomes assessment (Hawkins, Mathews, & Hamdan, 1999; Hersen & Rosqvist, 2008).

Suicidality and other forms of danger to self or other are so important that individualized assessment of these issues is normally conducted across as well as after treatment and at follow-up (Joiner, 2005; Rudd, 2006). The risks presented by suicidal thoughts and behavior are obvious, but there are a range of additional issues that need to be closely monitored given the risks they present to self or others such as eating-disordered behavior, high-risk sexual behavior, self-mutilation, the neglect or abuse of children or seniors, and substance abuse. Though some of these behaviors may be adequately assessed through the use of standardized, condition-specific measures, individualized assessment is often necessary to monitor progress or deterioration adequately.

Sources of Outcome Data

In Chapter 8, the importance of gaining information from the most reliable sources was emphasized as being critical for conducting accurate and useful intake assessments (e.g., the tripartite model by Strupp & Hadley, 1977). The same issue applies when evaluating the effectiveness of treatment. This is particularly true when the treatment goals involve the patient's functioning in social roles such as at work, home, or school or in the community. When treatment focuses on intrapersonal issues (e.g., depression or anxiety), it may be sufficient to gather outcome information only from the patient who may be a reliable reporter with regard to his or her internal state. When treatment focuses on externalizing disorders or the patient's functioning in social roles, however, patients may be unable or unwilling to provide reliable and thorough information regarding their behavior and performance. In many cases, for either intrapersonal or interpersonal issues, patients' perceptions of their behavior or performance varies significantly from that of family members, employers, educators, or various public officials (Miller & Berman, 1983). There may be no attempt to consciously or intentionally minimize problems, but individuals' perceptions and reports of their behavior and performance often need to be supplemented for the same reasons that information from additional sources is needed to conduct accurate and thorough intake assessments. Individuals are not always the most objective and reliable reporters with regard to their behavior and performance.

Engaging other significant individuals in the patient's life as collateral reporters on the patient's progress can also be helpful for maximizing social support and reinforcement for helping the patient to change, both during treatment and over the long term. In many cases, the likelihood of patient change increases as the amount of social support and engagement by significant individuals in patients' lives increases.

Schedule for Collecting Data

Therapists also need to decide on when to gather outcome data. Common practice is, at minimum, to assess patients at intake and then again at termination or discharge (Maruish, 2004b). It is very difficult to assess the effects of treatment without baseline data, so administering outcomes measures at the initiation of treatment is critical to conducting meaningful outcomes assessment. If therapists re-administer these instruments only at termination or discharge, they will be prevented from obtaining information from the significant number of patients who terminate treatment prematurely. There are also important advantages to monitoring treatment progress in a continuous manner, periodically throughout the treatment process (e.g., monthly; Sperry, Brill, Howard, & Grissom, 1996). This approach, known as *outcomes monitoring*, is very useful for identifying the significant number of patients who are not making treatment gains or who are actually deteriorating. Treatment can then be modified, consultations can be obtained or referrals made, or other changes can be implemented that may prevent treatment failures. For patients dealing with suicidality, it is critical that suicidal thoughts and behaviors are monitored continuously, normally at each contact with the patient (Joiner, 2005; Rudd, 2006).

Follow-Up

Maintaining treatment gains for months and years after treatment has ended is generally much more meaningful in patients' lives than making treatment gains that do not last beyond termination or discharge. Therefore, follow-up measurement of treatment outcomes after termination can provide the most important and meaningful measurement of the outcomes of therapy (Maruish, 2004b). Mailing or e-mailing patients at, for example, 3 or 6 months posttreatment can provide very useful follow-up data, though this request for information is likely to result in a low response rate when conducted by itself. A mailing can be followed up with a phone interview to the initial nonresponders to increase the response rate. Even when patients choose not to respond, it can be very important that they know that their therapist is concerned about their well-being and is available to offer support and treatment if needed. Posttreatment follow-up is especially important when the patient has dealt with more serious problems across the biopsychosocial domains. For example, when chronic suicidality is a concern (e.g., with individuals who have attempted suicide more than once), long-term monitoring of progress, recurrent treatment, and follow-up can all be critical to appropriate and quality care (Joiner, 2005; Linehan & Dexter-Mazza, 2008; Rudd, 2006).

The Biopsychosocial Approach to Outcomes Assessment

A health care-oriented biopsychosocial approach to conceptualizing behavioral health care in some ways requires the use of outcomes assessment. From this perspective, the primary purpose of behavioral health care is to meet patients' needs, and some type of outcomes assessment is necessary for evaluating how well these needs have been addressed.

A biopsychosocial approach to behavioral health care includes a broad focus that spans the biological, psychological, and sociocultural domains, addresses individuals' strengths in addition to problems and disorders, and incorporates a long-term developmental perspective as well. Depending on the focus and goals of treatment, assessment across all the biopsychosocial domains may be conducted at intake, during treatment planning and over the course of treatment, at termination, and also at follow-up. When dealing with cases involving high problem severity and complexity, routine practice may involve a combination of global, condition-specific, and individualized measures administered periodically across treatment and at follow-up. Though thoroughly monitoring outcomes in this manner is often unnecessary in cases of mild problem severity and complexity, a comprehensive holistic perspective on treatment outcome should be incorporated into the conceptualization of all types of practice. For example, most of the outcomes research on depression has relied on condition-specific measures such as the Beck Depression Inventory in the case of psychotherapy treatment studies, or the Hamilton Depression Rating Scale in the case of antidepressant medication research, measured at the beginning and the end of treatment (Fournier et al., 2010). Posttreatment follow-up measurement is infrequently conducted and the use of global measures is even less common. Given the high prevalence of chronic and co-occurring mental and physical health conditions as well as problems in the sociocultural domain, a broader and longer-term perspective on evaluating the effectiveness of treatment for depression would be very helpful in clinical research and practice.

The initial evaluation of the large-scale implementation of behavioral health care outcomes assessment in the United Kingdom suggests that systematic outcomes assessment can be highly useful. As mentioned in Chapter 9, the United Kingdom has undertaken a system-wide effort to improve access to behavioral health care. The program involves a graduated level-of-care model to address the full range of behavioral health issues through a wide variety of evidence-based practices (Clark et al., 2009). It also incorporates standardized outcome monitoring through the use of instruments such as the 34-item Clinical Outcomes in Routine Evaluation—Outcome Measure, the 9-item Patient Health Questionnaire Depression Scale, and the Patient Health Questionnaire Generalized Anxiety Disorder Scale (Barkham et al., 2001; Kroenke, Spitzer, & Williams, 2001; and Spitzer, Kroenke, Williams, & Lowe, 2006, respectively). This type of outcome assessment provides highly useful data to evaluate the effectiveness of treatment in individual cases as well as the national effort as a whole (e.g., outcomes measures at pilot sites have found very large treatment effect sizes, ranging from 0.98 to 1.26; Clark et al., 2009).

Another useful perspective for conceptualizing outcomes assessment from a biopsychosocial approach follows from the assessment model outlined in Chapter 8 that includes an assessment of problems and strengths across the 26 component areas of the biopsychosocial domains. Because a biopsychosocial approach to behavioral health care emphasizes the identification, utilization, and development of patient strengths and assets in addition to addressing problems and concerns, it can be informative to also measure the development of strengths as part of outcomes assessment. The case illustration below indicates how this model can be used to conceptualize change that occurs over the course of treatment across all the component areas. No standardized instruments are available for measuring these outcomes, but the model is useful for conceptualizing a comprehensive biopsychosocial approach to evaluating treatment effectiveness. Therapists' independently completed results could also be compared to patients' results, and those could be further compared to family members' results to gain a multifaceted assessment.

Case Example: Assessing Treatment Outcomes from a Biopsychosocial Approach with a Mildly Depressed Patient

The case example of a 49-year-old mildly depressed male patient was discussed in the past three chapters. This patient made significant improvement in several areas over the past 4 months since treatment began. The patient notes that he is very glad that he is getting back into physical shape. He now exercises regularly, his diet has significantly improved, and he has lost 15 pounds. He also has gotten back together with several old friends who still play basketball on a weekly basis. He reports "I feel like a new man—I feel like I'm 10 years younger." He reports having more energy and strength, and says, "I suppose it's obvious to you, but it really helps with a person's mood and general outlook on life."

In terms of the malaise he was feeling about his marriage, work, and family life, he reports that the examination of the effects of his parents' relationship and their approach to child rearing provided major insight into why he was feeling the way he was. He can now see that they provided a stable and reasonably healthy childhood for him and his siblings, especially when compared with how they were raised and what they learned about relationships and emotional well-being. He notes that his relationships with his mother, father, and siblings have recently actually improved somewhat: "Not a lot, but I'm surprised that they improved at all. If anything, we were all just steadily growing apart. But now, I'm actually hoping we'll continue getting closer to each other."

The patient reports that the most rewarding change has been in terms of his relationships with his wife and children. He says it was an excellent idea to have his wife join him for a few sessions, even though he was initially afraid how it would go and privately hoped she would decline the request. He says that he did not know that she was worried about him and she actually had some of the very same concerns as he did, though she was too afraid to admit them. They have started going out on dates again and "we're really enjoying each other's company, just like early in our marriage." Their sex life has returned as well. He says, "She won't admit it, but I think that my losing weight has increased my attractiveness to her again. Well, you've seen her—I mean, I wasn't exactly in her league." He is still worried that the boredom he felt about raising his children has caused them not to fully trust him. Though his relationships with them have improved, they still seem a little leery about his recently increased involvement in their lives. He says that he realizes it will take time before they can trust his motivations in this regard. He also reports that his examination of his feelings about his work has resulted in several important conversations with colleagues and friends who share his concerns about the ultimate value of their work. He does not yet know where these conversations might lead, but he is excited about several possibilities.

The treating psychologist then referred back to the original biopsychosocial assessment summary that she and the patient completed at the end of their first session and discussed the progress that was made across the biopsychosocial domains. After reflecting on the changes that occurred since his first visit and changes that he wants to solidify and continue to develop, the psychologist enters dots into the table (reproduced below) to indicate his level of functioning posttreatment as compared to pretreatment (indicated by checks; Table 11.1). Following this discussion, the psychologist and the patient decide to terminate treatment, though the patient says he will contact her if his progress reverses or questions arise. The psychologist also asks if he will complete the outcomes measures that he completed three times before. He quickly completes those, and they note that the scores indicate that his mood and functioning have returned to the normal or high range of functioning across all the areas assessed.

Additional Foci of Outcome Assessment

Comprehensive health care outcomes assessment serves additional purposes beyond questions regarding treatment effectiveness. Employers and managers are often very interested in patient satisfaction with the services provided by institutions, departments, or individual providers, and the cost-effectiveness of treatment is a critical concern as well.

Patient Satisfaction

Patients' satisfaction with health care services began receiving research attention in the 1950s, when it was noticed that increased patient satisfaction was associated with improved appointment keeping, medication use, and adherence to treatment recommendations (Williams, 1994). It was also found to be associated with a decreased likelihood of being sued for malpractice (Hickson et al., 1994). Patient satisfaction has recently grown in importance as the result of the increased marketing of health care services in the United States, and is also being given more attention as an indicator of quality of services. A variety of approaches to measuring patient satisfaction with services can be integrated relatively easily into clinic practice (see Smith, Schussler-Fiorenza, & Rockwood, 2006).

Cost Effectiveness

Concern about the cost-effectiveness of health care has been growing in the United States in recent years, particularly given the relatively poor health outcomes achieved while the amount spent on health care is by far the highest per capita in the world (World Health Organization, 2009). There is also an ethical concern that

Table 11.1 Pretreatment and Posttreatment Assessment of Biopsychosocial Functioning for the Case Example[a]

Biopsychosocial Domains and Components	−3 Severe need	−2 Moderate need	−1 Mild need	0 No need	+1 Mild strength	+2 Moderate strength	+3 Major strength
Biological							
General physical health							
Childhood health history							
Medications							
Health habits and behaviors							
Psychological							
Level of psychological functioning							
History of present problem							
Individual psychological history							
Substance use and abuse							
Suicidal ideation and risk assessment							
Effects of developmental history							
Childhood abuse and neglect							
Other psychological traumas							
Mental status examination							
Personality styles and characteristics							
Sociocultural							
Current relationships and social support							
Current living situation							
Family history							
Educational history							
Employment							
Financial resources							
Legal issues/crime							
Military history							
Activities of interest/hobbies							
Religion							
Spirituality							
Multicultural issues							

[a]Checks represent pretreatment and dots represent posttreatment.

society ought to spend its resources on the most effective health care services to benefit the largest number of individuals, and that one's ability to pay for those services is not a relevant issue for deciding who receives basic health care services. Many consider these to be moral obligations based on the principles of beneficence and justice (Beauchamp & Childress, 2009).

Cost-effectiveness analysis is the procedure used to determine whether health care treatments are beneficial to society in general (Nyman, 2006). The cost-effectiveness of treatments can be relatively easily examined in terms of number of lives saved or years of life gained as a result of providing particular treatments. Emphasis has shifted from these simple measures of treatment outcome to measures of quality of life because simply lengthening life if quality is not also achieved is not always preferable (Beauchamp & Childress, 2009). Quality-adjusted life-years (QALYs) have become the most widely used measure that attempts to combine the concept of quality of life with the quantitative extension of life to provide a more adequate measure of treatment outcome for use in cost-effectiveness analyses (Eddy, 1992). Integrating measures of quality of life into these analyses is similar to the concept of using generic measures of treatment outcome to measure more global aspects of patient functioning to supplement the information gained from using condition-specific measures.

Conclusions

Psychologists have always been very interested in the effectiveness of the treatment they provide. The most famous case studies in the field, such as Freud's discussion of the treatment of Dora, revolved around the outcomes of treatment. In the past, the outcomes of treatment for many types of psychotherapy were normally assessed through conversation during the termination stage of therapy. Patients and their therapists commonly reflected on the progress that was made along with challenges that could be anticipated and invitations to return to therapy as needed. More recently, adding standardized outcomes assessment to these conversations has been found to be highly valuable as well.

As discussed in Chapter 8, among the primary purposes of assessment are to provide ongoing monitoring during treatment as well as to conduct an outcome evaluation to assess the effectiveness of treatment. These are becoming more important priorities as demands for accountability grow and strategies are developed that can improve the quality of health care through the ongoing monitoring of responses to treatment. The potential for improving effectiveness through the monitoring of outcomes deserves special attention from clinicians and managers. Though outcomes assessment was easily avoided in clinical practice in the past, it is becoming increasingly unavoidable. This is particularly the case when behavioral health care is viewed from the biopsychosocial perspective advocated in this volume—the effectiveness with which patients' needs are met is a top priority from this perspective.

Most of the issues discussed in this chapter are now well researched, though some involve relatively new areas of investigation (e.g., the benefits of the ongoing monitoring of treatment progress). Consequently, psychologists need to follow the research to keep current on these issues. Nonetheless, the available literature on outcomes assessment suggests a number of best practices for performing outcomes assessment from the biopsychosocial health care orientation. A list of these practices would include the following suggestions:

- Collect baseline data to aid in interpreting results—outcomes cannot be fully interpreted without baseline data.
- For global and condition-specific measures, use standardized, normed, and psychometrically sound instruments so that the data obtained are more interpretable and normative comparisons are possible.
- Select global, condition-specific, and/or individualized outcome measures to assess the range of patient issues that are important in the individual case—outcomes assessment generally needs to be more comprehensive as problem severity and complexity increase.
- Focal or target complaints or symptoms are more likely to be measured with condition-specific or individualized measures than more global issues, which can be measured with generic instruments.
- Brief instruments that are quickly administered and scored are generally preferred—time and cost demands are important concerns in almost all types of clinical practice.
- Gaining information from multiple sources is often important, particularly when treatment focuses on externalizing issues or the patient's functioning in social roles (e.g., at home, work, or school or in the community).
- Administering multiple measurements over the course of treatment is very useful for monitoring the progress of treatment—it is very important to know if a patient is an early responder, is following the usual growth curve for patients with similar problems, is not benefiting, or is actually deteriorating.
- Follow-up is generally important but is especially important when patients have dealt with issues characterized by high rates of relapse or involving harm to self or others.
- Aggregate outcome data for quality improvement purposes. It is difficult to reliably identify strengths or areas for improvement without aggregated data; consequently, therapists should aggregate data from their individual practices and programs should aggregate at the program level.

Making full use of patient outcomes data has the potential to dramatically improve the effectiveness of behavioral and physical health care. The increased reliability and validity of outcomes measurement, along with more detailed and extensive therapist and treatment data, could lead to important behavioral health care improvements. These data can be aggregated on a large scale across large practice networks and provider systems to allow researchers to examine the characteristics of effective psychological treatment across therapists and for different types of patients with different disorders and circumstances. This will lead to a more thorough and detailed understanding of behavior change processes, more effective treatment, greater patient satisfaction with services, and improved cost-effectiveness. These are all priorities when behavioral health care is approached from a science-based, health care-oriented biopsychosocial approach.

Part IV

Additional Implications for Professional Psychology

The biopsychosocial approach to professional psychology has implications for practice, research, and education that extend beyond the treatment process. The next chapter discusses the importance of a preventive perspective on behavioral health. This perspective is critical if the health care professions are to improve their effectiveness at addressing the behavioral health needs and promoting the biopsychosocial functioning of the population in general.

The final chapter revisits the biopsychosocial metatheoretical framework in light of the issues raised in earlier chapters, and discusses the implications of the framework for research, practice, and education in professional psychology.

12 Prevention and Public Health Perspective on Behavioral Health

The biopsychosocial approach to professional psychology emphasizes meeting the behavioral health needs and promoting the biopsychosocial functioning of the general public. This obviously involves providing treatment for the behavioral health problems that individuals have already developed, but it also requires a preventive and public health perspective both for those who could be prevented from developing problems as well as for those who are showing early signs of developing problems. As is the case for medical health care providers, most professional psychologists will be employed providing treatment to those who have already developed problems. But the impact of both psychology and medicine will be seriously limited if providers are concerned only with treating problems after they have developed. Psychologists and physicians need to also support a public health and preventive perspective if they are going to significantly improve the health and well-being of the population in general.

The public health field takes a broad perspective on the health of the whole population. In addition to treatment, public health focuses on the etiology of health disorders, epidemiological surveillance of the health of the general population, disease prevention, health promotion, and access to and evaluation of the services provided to the public (Last & Wallace, 1992). Chapter 3 presented an epidemiological perspective to illustrate how professional psychology would be focused when a health care orientation is applied to meeting the behavioral and biopsychosocial needs of the general public. The present chapter adds to this public health perspective by focusing on prevention and health promotion. There are several important reasons why professional psychology needs to include these perspectives as priorities in professional psychology education and practice.

The Importance of a Public Health Perspective on Behavioral Health

The emphasis of professional psychology historically has clearly been on the assessment and treatment of individuals who have developed behavioral health problems and disorders, and not on the prevention of those problems. So it would be reasonable to ask why a chapter on prevention is included in a volume discussing the conceptual foundations of professional psychology. The public health field,

Foundations of Professional Psychology. DOI: 10.1016/B978-0-12-385079-9.00012-6

community psychology, and other specializations have a long history of important research, program development and evaluation, and advocacy on the prevention of behavioral health problems, while professional psychology has not. Nonetheless, there are several important reasons why professional psychology needs to increase the attention it gives to a public health perspective on behavioral health.

First, a public health perspective is needed to understand the context of behavioral health care. An adequate understanding of the behavioral health needs of the public cannot be obtained by focusing solely on treatment. Epidemiological, etiological, and preventive perspectives must also be applied to gain a broad understanding of the needs and circumstances of the public and the appropriate role of treatment providers in meeting those needs. For example, failing to appreciate the prevalence and co-occurrence of the full range of behavioral health and biopsychosocial issues that members of the public deal with will result in individual clinicians and the profession as a whole being less responsive, relevant, and effective, and consequently less valued as well. It is also critical to understand the etiology of behavioral health and biopsychosocial problems to work effectively when providing treatment to individuals, families, and communities. This is particularly important when taking a comprehensive biopsychosocial perspective where developmental, contextual, risk, and protective factors need to be considered to gain a full understanding of individuals' development and functioning. An orientation that does not include epidemiological, etiological, and preventive perspectives will provide a limited framework for conceptualizing professional practice in the field.

Second, it is simply more humane and ethical to prevent problems from developing, when it is possible and practical to do so, than it is to allow them to develop. As health care providers, our ethical obligations of beneficence compel us to attempt to prevent foreseeable problems from developing when possible and practicable. This is true of physical as well as behavioral health problems. Of course, professional psychologists provide treatment after problems develop, but treating those problems does not absolve us from the obligations as a profession to also work toward preventing problems from developing in the first place.

Third, behavioral health problems cause an enormous burden to the individuals affected, their families, and society generally, and it would be both efficient and humane to prevent as much of that burden as possible. In addition to having substantial effects on individuals and families, behavioral disorders are very costly in monetary terms. The direct costs of behavioral health services in the United States in 1996 were estimated to total $69 billion and an additional $12.6 billion for substance abuse treatment (US Department of Health and Human Services, 1999). Indirect costs such as lost productivity, Supplemental Security Income payments, welfare, homelessness, and incarceration are more difficult to estimate. Focusing only on lost income, Kessler and colleagues (2008) used data from the National Comorbidity Survey Replication and found that those with serious behavioral illness had personal incomes far below those without mental illness. The loss of personal earnings associated with serious behavioral illness for the nation as a whole was estimated to be $193.2 billion in 2002. Considering only the direct costs of treating mental illness and the personal income loss associated with mental illness, these two factors combined equal approximately 2% of the entire gross domestic product of the United States.

Using a measure of disease burden to estimate the years of life lost to premature death and years lived with a disability, the Global Burden of Disease study (Murray & Lopez, 1996) found that in established market economies such as the United States the burden of mental illness was second only to all cardiovascular conditions. In addition, the burden due to alcohol use was the fifth greatest factor and the burden due to drug use was the seventh greatest factor. Using data collected in the National Comorbidity Survey Replication, Merikangas and colleagues (2007) found that up to one third of illness-related days, when individuals were unable to carry out basic daily activities as usual, are related to mental rather than physical disorders. The largest number of days of role disability was attributed to musculo-skeletal disorders, followed by anxiety disorders and then mood disorders. All of these approaches find that the cost and burden of mental illness is enormous. Preventing these costs and burdens should be both a humanitarian and economic priority for the health care professions and for society in general.

A fourth reason that professional psychology should consider prevention to be a high priority is that it works. As will be seen later in this chapter, the effect sizes for prevention are reasonably strong and comparable to those of psychotherapy and medical interventions. There has been relatively little research investigating the cost effectiveness of preventive interventions, but the available evidence suggests they can be very cost effective as well. Durlak and Wells (1997) noted that preventive interventions focused on physical health promotion, early childhood education, and childhood injury were shown to save from $8 to over $45 for every dollar spent. Indeed, the landmark Institute of Medicine report on the prevention of behavioral disorders (1994, p. xvii) concluded that "There could be no wiser investment in our country than a commitment to foster the prevention of behavioral disorders and the promotion of behavioral health through rigorous research with the highest of methodological standards."

A fifth reason to make prevention a priority in professional psychology is that prevention is often the only effective approach to addressing certain problems. Several decades ago, the biologist Dubos made an impassioned appeal regarding the importance of prevention, arguing that "No major disease in the history of mankind has been conquered by therapists and rehabilitative modes alone, but ultimately only through prevention" (1959; in the same year, Albee made the same point with regard to behavioral health specifically). The modern evolution of medicine has led to major gains in the ability to cure individual cases of disease, but its overall impact on morbidity and mortality has been minimal (McKeown, 1979). In contrast, public health measures based on a more complex biopsychosocial understanding of disease and relationships among individuals and community life has had a major impact on morbidity and mortality (McNeill, 1979).

It appears that no widespread disease or disorder has ever been controlled or eliminated through the individual treatment of those who had the disease or disorder (Albee, 2006). Given that more than 25% of Americans have a behavioral health disorder in any given year and 50% have one or more disorders over the course of their lifetimes (Kessler et al., 2005), the resources that would be needed to treat all those individuals would be tremendous. And even if those resources could be made available, it is unlikely that behavioral disorders would be controlled or eliminated as a result.

A public health perspective is critical to a comprehensive understanding of behavioral as well as physical health, particularly when applying a biopsychosocial framework to conceptualizing professional psychology. There are highly important ethical, practical, and economic reasons why more attention should be given to this perspective. To provide an introduction to the topic of behavioral health prevention, the following sections summarize essential concepts used in the field and the evidence regarding the effectiveness of preventive interventions for behavioral health problems.

Basic Concepts

The principles of prevention were first applied to the control of infectious diseases through the use of mass vaccination, water safety, and other public health measures, and were later applied to chronic diseases and behavioral disorders (Institute of Medicine, 1994). The prevention of chronic diseases and behavioral disorders is typically more complicated than preventing infectious diseases, however, because the latter have specific, precise causes while the causes of chronic and behavioral disorders tend to be far more multifactorial and complex. In both cases, however, prevention can be highly effective.

Definitions of the general categories of public health prevention were developed over a half century ago (Commission on Chronic Illness, 1957) and continue to be the primary definitions used in the field. Primary prevention refers to the prevention of a disease before it occurs; secondary prevention refers to the prevention of recurrences or exacerbations of a disease that has already been diagnosed; and tertiary prevention refers to the attempt to reduce the amount of disability that is caused by a disease.

These definitions work less well in the case of behavioral health, however, so the Institute of Medicine in 1994 suggested modified definitions of the different levels and forms of prevention. In this approach, the definition of *primary prevention* is essentially the same as the earlier definition, referring to interventions that help prevent the initial onset of behavioral disorders. Instead of secondary prevention, however, *treatment* is used to refer to the identification and treatment of individuals with behavioral disorders. Finally, *maintenance* is used to refer to interventions designed to reduce relapse and to provide rehabilitation for those with behavioral disorders. The Institute of Medicine further distinguished between *universal*, *selective*, and *indicated* preventive interventions. These are distinguished by the population groups that are targeted for intervention, namely (1) the population in general, (2) groups at greater than average risk for developing problems, or (3) groups identified through a screening process designed to classify individuals

Table 12.1 Basic Types of Prevention Strategies and Interventions

Commission on Chronic Illness (1957) definitions:
- *Primary Prevention*—interventions to prevent a disease before it occurs
- *Secondary Prevention*—interventions to prevent recurrences or exacerbations of a disease that has already been diagnosed
- *Tertiary Prevention*—interventions to reduce the amount of disability caused by a disease

Institute of Medicine (1994) definitions applied to behavioral health:
- *Primary Prevention*—same as above: interventions to prevent a disease before it occurs
- *Treatment*—the identification and treatment of individuals with behavioral disorders
- *Maintenance*—interventions to reduce relapse and to provide rehabilitation for those with behavioral disorders

Types of primary preventive interventions based on target group:
- *Universal*—the population in general
- *Selective*—groups at greater than average risk for developing problems
- *Indicated*—groups identified through a screening process designed to classify individuals with early signs of a problem

who exhibit early signs of the problem. The SBIRT model discussed in Chapter 9 to identify and assist those at risk for substance abuse (the SBIRT acronym stands for screen, brief intervention, brief treatment, and referral for treatment) is an example of this last type of preventive intervention. These different types of prevention strategies and interventions are summarized in Table 12.1.

Health promotion programs often share the same ultimate goal as prevention programs—to increase the likelihood that individuals avoid maladjustment—but the approach to reaching that goal is different. Prevention focuses on avoiding risk factors, whereas health promotion programs aim to develop competency and promote wellness (Cowen, 1994). Health promotion aims to identify and strengthen protective factors such as supportive family, school, and community environments that enhance well-being and help children and adults avoid adverse emotions and behaviors (National Research Council and Institute of Medicine, 2009). As competence and psychological well-being improve, resilience increases and individuals are better able to respond to stressors and influences that might otherwise lead to maladjustment.

Risk and Protective Factors

The identification of risk and protective factors are critical aspects of prevention research. Risk factors are "characteristics, variables, or hazards that, if present for a given individual, make it more likely that this individual, rather than someone selected from the general population, will develop a disorder" (Institute of Medicine, 1994, p. 6). Some risk factors are generally not malleable to change, such as genetic inheritance or gender, while other risk factors are relatively easily changed, such as lack of social support, low reading ability, or being victimized by bullying (e.g., Durlak & Wells, 1997). Even in the case of disorders with very high heritability, modifying the environment is sometimes highly effective at reducing the risk that the disorder will develop. For example, phenylketonuria is one of the few genetic disorders that can be very effectively controlled in this way. Unidentified and untreated, phenylketonuria can lead to serious irreversible brain damage, including behavioral retardation and seizures, but very early detection and avoiding foods high in phenylalanine (e.g., breast milk, dairy products) can be completely successful for avoiding these problems (Centerwall & Centerwall, 2000). Most behavioral disorders, however, have a far more complex etiology, and identifying the causal risk factors involved is much more complicated.

Protective factors are internal or external influences that improve an individual's response to a risk factor (Rutter, 1979). Supportive parents or other adults from the community, for example, are important protective factors against the development of a wide range of maladaptive outcomes in children (National Research Council and Institute of Medicine, 2009). Resilience has also received a great deal of support as an internal protective factor affecting an individual's response to stressors or traumatic events (Garmezy & Rutter, 1983).

Prevention efforts are often aimed at risk reduction (Institute of Medicine, 1994). After research has clarified the interaction of risk and protective factors, investigation generally turns to identifying the causal risk factors that are malleable and potentially alterable through intervention. Once these factors have been identified, preventive interventions are designed and the effects of the interventions can then be evaluated, often through preventive intervention trials. The prevention of physical disorders is also often similar to the prevention of behavioral disorders in that both can have complicated multifactorial causes. For example, various risk and protective factors are related to the development of both cardiovascular disease and substance abuse. Interventions aimed at reducing risks and enhancing protective factors (e.g., smoking cessation, changing diet, and increasing physical activity) can be quite effective at reducing risks for morbidity and mortality caused by cardiovascular disease (e.g., Flora, Maccoby, & Farquhar, 1989), while reducing risks and strengthening protective factors related to the family, school, and community reduce the likelihood that young people will abuse substances (National Research Council and Institute of Medicine, 2009).

A preventive perspective is also very useful when providing treatment or maintenance interventions. For example, when a clinician is providing maintenance interventions for patients with chronic conditions, it is important to identify risks for relapse or deterioration for that particular individual. This strategy is typically integrated into the treatment of substance dependence, where it is important to identify the risky environments, thoughts, or feelings that can trigger relapse in the individual case. This risk reduction intervention occurs within the context of treating an individual patient as opposed to preventing the onset or worsening of substance abuse in the population in general, though both strategies involve the use of risk reduction principles (US Department of Health and Human Services, 1999).

Many behavioral health problems share some of the same risk factors. Low birth weight, for example, is a risk factor for many maladaptive outcomes, but when it exists in the presence of social risk factors, the chances of negative outcomes increase significantly (McGauhey, Starfield, Alexander, & Ensminger, 1991). Rutter (1979) first proposed the cumulative risk model to show that as the number of risks that children face increases, the developmental status of the child decreases. Children who have difficult temperaments and low intelligence, who live in families with serious parental conflict, violence, substance abuse, or behavioral disorder, and who live in a distressed community with inadequate schools face an accumulation of risk factors associated with a variety of negative outcomes. In an examination of a national probability sample of children up to 3 years of age investigated for child maltreatment, Barth and colleagues (2008) found that 5% of children exposed to no or one risk factor in addition to maltreatment had a measurable developmental delay in their cognitive, language, or emotional development. As the number of additional risk factors increased (e.g., minority status, poverty, single caregiver, domestic violence, caregiver substance abuse, caregiver mental health problem, low caregiver education), the proportions with developmental delays quickly rose as well. Of those exposed to four risk factors, 44% had developmental delays; of those exposed to five risk factors, 76% had developmental delays; and of those exposed to seven risk factors, 99% had measurable developmental delays.

Likewise, neuroticism as a personality characteristic has been associated with a wide variety of Axis I and II disorders, serious physical health problems, shorter life span, and several indicators of poorer quality of life, such as marital dissatisfaction and lower occupational success (Lahey, 2009; Malouff, Thorsteinsson, & Schutte, 2005; Saulsman & Page, 2004). Neuroticism in parents has also been found to be associated with behavioral and emotional problems in their children that persist and lead to chronic interpersonal difficulties in adulthood (Ellenbogen, Ostiguy, & Hodgins, 2010).

Because some risk and protective factors are more malleable and amenable to intervention than others, Robins (1970) advocated that preventive interventions should aim at "breaking the chain at its weakest links." For example, it may be easier to improve the behavior and academic achievement of a child who comes from a family with severe parental conflict and substance abuse than it is to attempt interventions aimed at changing the parents' behavior (US Department of Health and Human Services, 1999).

Effectiveness of Preventive Interventions

Just as with the case of psychotherapy, the issue of the effectiveness of preventive interventions for reducing behavioral disorders or changing maladaptive behavior was highly controversial for many years. In fact, there was a great deal of skepticism about the effectiveness of prevention interventions until the late 1990s. And like the case of psychotherapy, it was meta-analysis that provided the compelling evidence that finally answered this question.

The first large-scale meta-analysis of controlled outcome studies of preventive interventions was conducted by Durlak and Wells (1997). They examined 177 studies from 1991 or earlier that (1) were designed to reduce the incidence of adjustment problems and promote behavioral health; (2) included children and youth younger than age 18; (3) included a control condition; and (4) focused on behavioral and social functioning. Various categories of prevention programs were examined, including parent training, divorce adjustment, school adjustment, awareness and expression of emotions, and interpersonal problem-solving skills. The interventions included both prevention and health promotion programs. The meta-analysis found that all categories of programs were associated with positive effects, and the mean effect sizes ranged from 0.24 to 0.93. These effect sizes are similar to, and often higher than, those achieved by many educational, psychotherapeutic, or medical interventions. Several meta-analyses conducted since then have found that preventive interventions are effective with regard to diverse problem behaviors such as drug use in adolescents (Tobler & Stratton, 1997), child sexual abuse (Davis & Gidycz, 2000), and depression in children, adolescents, adults, and seniors (Jane-Llopis, Hosman, Jenkins, & Anderson, 2003; see also the National Research Council and Institute of Medicine, 2009). A recent meta-analysis found that early developmental prevention programs for children 0–5 years of age were associated with a variety of improvements in later adolescent functioning, including greater educational success and reduced criminal behavior (Manning, Homel, & Smith, 2010).

In addition to preventing behavioral health problems, prevention and health promotion have the potential to greatly reduce medical illness and the need and demand for medical services as well. For example, it is estimated that relatively inexpensive primary prevention and health promotion interventions can prevent up to 70% of the world's global disease burden (i.e., the disease, disability, and death suffered by the global population as a whole—Fries et al., 1993; World Health Organization, 2002). There is great potential for prevention to reduce both behavioral and physical health problems in the United States as well as globally.

Conclusions

Public health measures such as sanitary sewer systems and vaccines have been tremendously effective in improving the physical health of the general population (McNeill, 1979). Indeed, these measures and the resulting improvements in morbidity and mortality have simply been transformative in the countries where they have been applied. There have been relatively few widespread efforts to prevent psychological dysfunction and promote behavioral health, however. The sociological and political reasons for this are complicated and extend beyond the scope of this chapter (for perspectives on this issue, see Prilleltensky, 2008). The ethical, practical, and economic reasons to prevent maladjustment and promote behavioral health nonetheless remain.

There was a great deal of excitement about the potential of a public health approach to behavioral health in the late 1980s and early 1990s as prevention research advanced and behavioral health prevention became a national priority. Three milestone reports reflected the growing importance of behavioral health prevention for the national agenda in the United States at that time (*Prevention of Behavioral Disorders: A National Research Agenda* by the National Institute of Behavioral Health in 1993; *Reducing Risks for Behavioral Disorders: Frontiers for Preventive Intervention Research* by the Institute of Medicine in 1994; and *A Plan for Prevention Research for the National Institute of Behavioral Health* by NIMH in 1995).

This excitement soon abated, however, as prevention became a lower priority for the federal government in the second half of the 1990s (Mrazek & Hall, 1997). Federal funding for prevention research stalled and the recommendations of the above landmark reports were not implemented. Evidence regarding the effectiveness of preventive interventions continued to accumulate, however, and moral support was provided by the New Freedom Commission on Mental Health (2003). There was also growing appreciation of the impact of behavior on physical disease and health as well. The Institute of Medicine (2004) concluded that roughly 50% of morbidity and mortality in the United States is caused by behavior and lifestyle factors, and that medical education needs to strengthen training in behavioral impacts on health if physicians are going to increase their effectiveness at treating patients' physical health problems.

The importance of prevention for both behavioral and physical health is quite clear. Making it a priority for governments, institutions, and society, however, is much more complicated. The conclusions reached by George Albee, the best known researcher and advocate in the behavioral health prevention field (Britner, 2007), present a strong case for making prevention a higher priority. Two years before he died in 2006, Albee addressed the Third World Conference on the Promotion of Mental Health and Prevention of Mental and Behavioral Disorders and presented five general conclusions that he believed were justified by the available research:

1. "One-on-one treatment, while humane, cannot reduce the rate of behavioral disorders . . .
2. Only primary prevention, which includes strengthening resistance, can reduce the rate of disorders. Positive infant and childhood experience are crucial. Reducing poverty and sexism are urgent strategies.
3. Ensuring that each child is welcomed into life with good nutrition, a supportive family, good education, and economic security will greatly reduce emotional distress . . .
4. Cultural differences in diagnoses must be understood and be part of program planning.
5. Strong differences of opinions about causes—particularly brain disease versus social injustice—must be resolved by unbiased scientific judgment before real progress can be made" (Albee, 2006, p. 455).

Ethical and scientific perspectives on the health of the general population support giving high priority to the prevention of psychological maladjustment and the promotion of mental health and biopsychosocial functioning in general. From a practical perspective, it appears that it is not possible to significantly lower the high prevalence and massive burden of behavioral health problems without more emphasis being given to prevention. Even though most professional psychologists will gain employment by providing behavioral health care treatment and maintenance interventions, the profession as a whole needs to advocate more strongly for a preventive perspective on the behavioral health of the general population. The findings of Keyes (2007, discussed in Chapter 3) highlight the necessity of incorporating this perspective. He found that only 2 in 10 Americans are "flourishing" (a state of positive mental health and life satisfaction) and, at the other end of the continuum, 2 in 10 are "languishing" (a state of poor mental health). The large majority in between tend to experience significant functional impairment and significant physical and mental health problems. The proportion who are flourishing should be far larger, and the proportion who are languishing should be much smaller. The size of this problem simply requires prevention and health promotion interventions—major improvements in this situation will not result from treatment and maintenance interventions alone.

The biopsychosocial approach to conceptualizing behavioral health emphasizes a comprehensive, developmental perspective on human development, functioning, and behavior change. The interaction of biological, sociocultural, and psychological influences and the role of risk and protective factors are essential aspects of this approach. As a result, a preventive perspective on behavioral health is unavoidable. A comprehensive understanding of the development and functioning of human beings in general, as well as particular individuals, requires the integration of epidemiological, etiological, preventive, and health promotion perspectives.

Professional psychologists should take more responsibility for preventing behavioral health disorders and promoting positive mental health. They should also take more responsibility for prevention and health promotion with regard to medical and sociocultural functioning in general. Given that human development and functioning are highly sensitive to and dependent on the interaction of biological, psychological, and sociocultural influences, psychologists are well positioned to take a leading role in promoting biopsychosocial health and functioning. At a practical level, this will require that they become more interdisciplinary in perspective and work more collaboratively with other human service fields. Fortunately, a wide variety of health care and human service professions share with psychology the commitment to promoting the health of the whole person and the whole population.

13 Conclusions and Implications for Professional Psychology Education, Practice, and Research

Both the scientific and applied areas within the discipline of psychology have grown dramatically over the relatively short history of the field, and professional psychology in particular has grown very rapidly in size and influence over the last half of that period. Psychologists provided relatively little mental health care before World War II but soon overtook psychiatry as the primary provider of psychotherapy services. The first psychologists became licensed to practice in 1945, and there are now over 85,000 licensed psychologists in the United States (Duffy et al., 2006). Behavioral health care in the United States has been transformed as a result.

While it was experiencing dramatic growth and development, psychology also endured major conflict and contention between schools of thought and the various theoretical camps. There has been substantial contention and divisiveness between scientists and practitioners and between adherents of different theoretical, methodological, and educational orientations. There have been few times when the field has been able to address challenges or undertake initiatives collaboratively as a whole. As discussed in Chapter 4, competition between schools and camps is natural for a young science—all sciences go through a similarly contentious period when they are getting established. In addition, psychology examines the most complicated of natural phenomena. It was inevitable that there would be competition and conflict between alternative perspectives on the best ways to understand psychological phenomena.

The basic premise of this book, however, is that the field is now ready to leave behind the pre-paradigmatic era of conflicting theoretical orientations to understanding professional psychology and replace them with a unified science-based framework. It was argued in Chapter 1 that there are two critical issues that need to be resolved before the field can leave behind its conflictual past and move forward under a unified conceptual framework. The first issue concerns the lack of clarity in the definition of professional psychology such that the nature, scope, and purposes of the profession are not clearly identified. The second issue concerns the scientific basis for professional psychology. Though a great deal is not yet known regarding many psychological processes, is the amount that is known sufficient to justify a general transition away from the traditional theoretical orientations used in the field to a unified science-based biopsychosocial framework for understanding

Foundations of Professional Psychology. DOI: 10.1016/B978-0-12-385079-9.00013-8

and practicing psychology? Resolving these two issues will have far-reaching impacts for the next steps in the evolution of the field.

Conclusions Regarding the Two Critical Issues Needing Resolution in Professional Psychology

Without resolution of these two fundamental questions regarding the basic nature of professional psychology, it will be difficult for the field to come together around a shared, unified conceptual framework, perspective, and sense of purpose. These two issues are fundamental to the nature, identity, purpose, and role of the profession, both for practitioners within the field as well as for stakeholders outside. Reaching consensus regarding these fundamental issues may be the impetus that allows the field to leave behind its conflicting pre-paradigmatic past and enter a more focused, less conflictual, and potentially more efficient and effective paradigmatic era in the development of the profession.

There are several signs of movement on these issues. With regard to the first question, the discussion in Chapter 2 noted that there has long been confusion regarding the nature, scope, and purpose of professional psychology, and consequently also the core knowledge, skills, and competencies that are needed to practice psychology. Given the complicated historical development of professional psychology within the discipline of psychology and within the broad mental health care field that was dominated by psychiatrists (Benjamin, 2007; Grob, 1995), the lack of progress in developing a clear definition of the field was understandable.

The field now appears ready to move beyond that stage. A wide variety of professional and governmental organizations have recognized professional psychology as a health care profession, and states license psychologists and health insurance companies and governmental agencies pay psychologists in their role as health care providers. Professional psychologists obviously provide several types of services other than behavioral health care (e.g., forensic psychology, executive coaching, sports psychology), but relatively few make their living by providing these other services. The APA is also increasingly recognizing the role of psychology as a health care profession (APA Presidential Task Force, 2009; Johnson, 2001).

The definition of professional psychology provided in Chapter 2 is centered around a health care orientation that is founded squarely on science and ethics. Also integrated into the definition is a biopsychosocial perspective on psychology and health care. Indeed, it was argued that a biopsychosocial perspective is necessary for a scientific understanding of human psychology (Chapters 4 and 5), for the appropriate application of that understanding in health care (Chapter 7), and even to gain a comprehensive understanding of ethics and moral behavior (Chapter 6). Chapter 3 illustrated how this definition can be used to clarify the range of knowledge and skills needed to address the behavioral health and biopsychosocial needs of the general population, and Chapters 8–11 illustrated how this conceptualization can be used to inform the major phases of the treatment process. If this type of definition is successful in clarifying the nature, scope, and purposes of the field, it would help bring psychologists together around a unified basic perspective that could go a long way toward resolving many of the historical controversies that have long divided the field.

There are also signs that consensus is emerging regarding the second major question for the field which concerns the theoretical and scientific underpinnings of professional psychology. Though there are clear historical reasons for the complicated and confusing conceptual foundations of the field (see Chapter 4), there are also clear signs that the field is moving toward a unified, science-based biopsychosocial metatheoretical framework for understanding human psychology. Detailed explanations of complex psychological processes are far from complete, but the scientific tools now available to investigate these questions have evolved dramatically in recent decades and more complete explanations are emerging for many psychological phenomena.

At some point, the scientific understanding of human psychology will be sufficiently comprehensive and detailed to be able to provide a unified conceptual framework for understanding human development, functioning, and behavior change. When that point is reached (if it has not been already), a single unified scientific framework will replace the collection of theoretical orientations that have historically been used to conceptualize these processes. The systematic adoption of such a framework will resolve many of the conflicts that resulted when competing theoretical orientations were used to conceptualize the nature of human psychology and clinical practice in the field.

That point may now have been reached. The scientific understanding of psychology has progressed significantly in recent years, and there is far less controversy regarding recent research findings explaining different aspects of human development, functioning, and behavior change. Deciding when the science is strong enough to justify a general transition away from traditional practices is complicated, however, because this is a "tipping point" decision, not an "all or nothing" decision point. If professional psychology must have thorough and detailed explanations of human psychology before behavioral health care can be based on a unified science-based theoretical framework, then that point has not been reached. But that is not the question that needs to be answered. Biology, chemistry, and physics do not yet have complete explanations for many natural phenomena, and yet society is quite comfortable relying on the incomplete knowledge that is available in those fields for informing medicine, engineering, and the many other applied areas that we are so heavily dependent on in modern life. No one would argue that we must wait until detailed and complete explanations of all biological and physical phenomena are available before we can safely rely on medical interventions and engineering technology. When scientific and applied fields are practiced in a responsible and conscientious manner, it is possible for medicine and engineering to be practiced safely and with great effectiveness without the science behind them being complete. The question for professional psychology is similar. Are the currently available scientific explanations of human psychology sufficient to justify a general transition to a unified science-based metatheoretical framework for the field?

This is obviously a very important question for the field. Consequently, it is useful to examine the larger context of contemporary psychological science when considering this question. Highlighting recent advances in the scientific understanding of human psychology helps clarify the question of whether the "tipping point" has been reached.

Professional Psychology in the Midst of Remarkable Scientific Progress

Science has been progressing at a truly remarkable pace in recent years. It will take time before scientists and historians are able to look back and assess whether the present period represents a time of revolutionary progress. But from the perspective of the present, the progress currently being realized is certainly remarkable. From the level of particle physics through the biological and behavioral sciences and all the way to astrophysics, new tools and data analytic capabilities are allowing researchers to tackle questions that could not be examined directly before. In the biological and psychological sciences in particular, for the first time scientists are able to investigate the human mind and brain in ways that begin to capture the tremendous complexity that was always evident but was simply beyond the reach of experimental methods. A steering group for the National Science Foundation (Wood et al., 2006) reflected on developments in the neurosciences in particular and drew the following conclusions:

> "Science now stands at a special moment in humankind's long history of thinking about the brain, a moment of revolutionary change in the kinds of questions that can be asked and the kinds of answers that can be achieved ... In the past, experiments typically focused on a single type of molecule in the brain, the electrical activity of a single neuron, or the connections from one cell to the next. Advances in chemistry, molecular biology, physics and engineering have allowed scientists to move beyond this 'one at a time' approach. Thus it is progressively becoming possible to catalog all the molecules involved in a particular signaling pathway, to record the activity of hundreds of neurons simultaneously, or to diagram a complex neural circuit completely ... Rather than investigating limited sets of proscribed behaviors, new high-resolution measurement techniques make it possible to investigate complex behaviors over long periods of time as they occur naturally and spontaneously ... The integration of these approaches offers the hope of a truly predictive theory of the brain" (p. 6).

In the past, psychology had to rely on either "bottom-up" or "top-down" research strategies that examined relatively small expanses of the tremendously complex human nervous system. Researchers could explore phenomena at just one or a few organizational levels at a time. As a result, investigations often focused on either bottom-up biologically oriented explanations of the function of basic neuronal systems (e.g., in the giant marine snail, *Aplysia*; Kandel, 2006) or top-down psychological explanations at high levels of organization, such as psychopathology, personality, and intelligence (e.g., through factor analytic studies of psychological tests). The research methodologies needed to span these two extremes were not available.

For the first time, however, scientists are now beginning to be able to examine and build real-time working models of complete neuronal systems. For example, Watts (2003, p. 3) described how he and his colleagues used "reverse engineering of the brain" to build a detailed replica of part of the human auditory processing system that could replicate the truly remarkable properties of audition such as the ability to identify and locate sounds in the midst of significant background noise (the "cocktail party effect"). This research has led to the development of speech recognition programs that are reaching high levels of accuracy even in noisy conditions. Researchers have also copied the neuronal organization and function of all five cellular layers of the retina, making it possible to emulate the visual messages that the retina sends to the brain (Boahen, 2005). Inefficient but functional retinal prostheses are already providing blind individuals with limited vision and the hope that high-fidelity prostheses providing near-normal vision will soon be available.

A prominent example of the remarkable pace of recent scientific discoveries is the human genome project. The remarkable speed with which the human genome was read and sequenced could not have been predicted even just 20 years ago— indeed, the project was completed 2 years ahead of schedule (Collins, Green, Guttmacher, & Guyer, 2003). It was the ability of computer science concepts and tools in conjunction with quickly increasing computing capacities that allowed scientists to develop effective mathematical abstractions for describing findings. These DNA sequence databases were easily shared, compared, critiqued, and corrected, and consequently consensus findings emerged quickly and efficiently (Emmott & Rison, 2006).

Even the quintessentially human characteristic of consciousness is beginning to be investigated at the level of the neural mechanisms involved. The human prefrontal cortex is the most greatly expanded area of cortex compared with other mammals and is critical to humans' uniquely capacious intellectual abilities. O'Reilly (2006), for example, has found that the prefrontal cortex is able to plan and carry out complex goals and plans through a synthesis of analog and digital forms of computation, and that "perhaps a fuller understanding of this synthesis of analog and digital computation will finally unlock the mysteries of human intelligence" (p. 94).

Examples of cutting-edge research that are providing revolutionary new perspectives on understanding natural phenomena seem to be emerging almost daily. Of course, the current pace of research progress may not continue. There is no doubt, however, that scientific knowledge in many areas, particularly with regard to the human mind and brain, is advancing at breathtaking speed. Indeed, observers have frequently noted that scientific knowledge in general is advancing at an exponential rate (e.g., Hawkins, 1983; Kurzweil, 2005). Gordon Moore, one of the inventors of integrated circuits, observed in 1965 that the computational capacity of computers based on the shrinking size of transistors on an integrated circuit was doubling every year. This exponential rate of growth, now referred to as "Moore's Law," continues to be realized and has been a major factor in propelling the extraordinary pace of scientific and technological progress in recent decades.

Given the tremendous complexity of human psychology, the scientific understanding of the complex psychological processes that are often the focus of behavioral health treatment was limited until recently. In the past, psychologists had no alternative but to rely on one or more of the available theoretical orientations for conceptualizing clinical cases, even though these competing theories were widely considered to be incomplete. Advocates of competing theories often believed the opposing theories were hopelessly flawed, and it was quite unclear which of the many available theories might be the best. Nonetheless, they were the best available, and a selection needed to be made if one's clinical interventions were going to have coherence and consistency. Great progress has been made in understanding the mind and brain, however, and the question now is whether the assortment of competing theoretical orientations traditionally used to conceptualize psychological practice should be replaced with a unified science-based theoretical framework.

Though scientific knowledge regarding many psychological processes is still far from complete, a strong argument can be made that current scientific knowledge is now sufficient to support a unified biopsychosocial metatheoretical framework for understanding human psychology and behavioral health care. In fact, the tipping point in support of this argument may have been reached several years ago already—the biopsychosocial framework has been mentioned so frequently across psychology in the past few years that it appears to be the *de facto* paradigmatic framework for the field. Indeed, it is hard to detect any significant disagreement on this view—it may now be essentially unanimous that human development and functioning cannot be understood without an integrative, holistic approach that spans the biopsychosocial domains. Explaining the functioning of the entire system is tremendously complex, of course, but the general metatheoretical perspective that is required now appears clear.

Endorsing a unified biopsychosocial approach to professional psychology in an intentional and systematic manner would allow the field to leave behind its conflictual pre-paradigmatic past and move ahead with a unified perspective, effort, and purpose. This could result in the field becoming much more efficient and effective, but it would also require several changes in educational, research, and clinical practices that could be quite challenging. Previous chapters noted many of these changes at various points, whereas the section below consolidates and summarizes the discussion of changes that would likely result from the widespread adoption of this framework.

Implications for Education and Licensure

The impact of taking a unified biopsychosocial approach to professional psychology education would be substantial. Focusing on meeting the behavioral health and biopsychosocial needs of the general public through a conceptual framework founded on science and ethics would result in several significant changes to the traditional curriculum that has been used in many professional psychology education programs.

For many psychologists, perhaps the most significant result of transitioning to a unified biopsychosocial approach to professional psychology would involve reconceptualizing the role of the traditional theoretical orientations. Instead of the traditional approach to learning the profession, which involves selecting and mastering a theoretical orientation to guide one's clinical practice, students would learn to conceptualize cases in a holistic biopsychosocial manner. After a comprehensive, integrative psychological assessment is completed, then treatment plans can be developed that incorporate the variety of therapies that have been found to be effective in achieving behavior change. From the perspective of the biopsychosocial approach, the traditional theoretical orientations would be largely reconceptualized as therapies that well-controlled research have found to produce therapeutic benefit.

Another likely result of transitioning to a biopsychosocial framework for professional psychology would be the incorporation of an epidemiological perspective for understanding the behavioral health needs and biopsychosocial functioning of the general public. As noted in Chapter 3, the most common psychiatric conditions and concerns in the general public involve sexual dysfunction and substance dependence (nicotine and alcohol). In addition, nearly 50% of the population deals with chronic medical conditions. Issues related to child maltreatment, educational and vocational achievement, relationship and family dysfunction, and legal difficulties and crime are also highly prevalent, and all are known to have a significant impact on development and functioning. Culture, religion, spirituality, sexual orientation, and gender are all important as well. Several of these issues are covered in only a cursory fashion in many professional psychology training programs, and sometimes they are not covered to any significant extent at all. These factors are clearly important in terms of both their developmental influence and their impact on patients' current functioning and treatment. Taking a biopsychosocial approach to understanding development, functioning, and behavior change would result in a significant broadening and deepening of the curriculum across the sociocultural, psychological, and biological domains.

The health care orientation of the biopsychosocial approach also focuses attention on the effectiveness of treatment, and this has two important implications for professional psychology education. One involves the emphasis of the evidence-based practice movement on identifying therapeutic methods and therapies that are effective for treating behavioral health problems. This also requires sufficient training in statistics, measurement, and research design to be able to keep current with the research literature. This training would emphasize methods used to evaluate clinical research such as psychometrics, test sensitivity and selectivity, the methodology of clinical trials and meta-analysis, and different approaches to measuring effect size and treatment effectiveness. A second implication of the attention to treatment effectiveness would be the focus on using outcomes assessment to evaluate and improve the effectiveness of treatment in individual clinical cases.

This topic currently receives less emphasis in discussions of evidence-based practice but is critical in the movement toward accountability in health care. These two approaches can also be used to better inform each other. Systematically gathered outcomes assessment data from actual clinical practice can be aggregated and examined to learn more about the nature of effective treatments (e.g., different therapies and therapists for different kinds of patients with different problems and biopsychosocial circumstances). This in turn can lead to more useful guidelines for intake assessment, treatment monitoring, and outcomes assessment.

Procedures for assessing student competence to practice professional psychology would be significantly affected as well by transitioning to a biopsychosocial framework. Assessing student competence currently is very difficult because students commonly choose from a variety of theoretical orientations to guide their professional practice. Current APA Commission on Accreditation guidelines, internship application procedures, and state licensure and other guidelines all generally allow substantial latitude in the approach one can take to clinical practice and typically leave it to the individual student or practitioner to decide what his or her theoretical orientation and approach to practice will be. As a result, one student may develop competence to conceptualize cases and implement interventions from the perspective of a particular theoretical orientation, while another student using a different orientation may develop a different set of competencies. And an independent evaluator operating from another orientation might judge both of their approaches to be inadequate. Taking this splintered approach to identifying professional competencies makes it extremely difficult to assess the attainment of competence in a reliable and valid manner. A unified biopsychosocial approach, on the other hand, provides a common perspective for identifying the knowledge and skills needed to assess and treat the behavioral health and biopsychosocial needs of patients.

A question that has not received much research attention to date involves the range of behavioral health care issues that psychologists should be competent to address in clinical practice. Specifically, what range of disorders and issues should psychologists be able to diagnose and assess in what range of demographic and diagnostic populations? And second, what range of disorders and issues should psychologists be able to then treat? (In general, health care providers are able to diagnose, at least provisionally, many more conditions than they are able to competently treat.) Too great a scope of practice would obviously be unmanageable, though it is notable that the generalist field of family medicine takes an expansive approach to this question. The American Academy of Family Physicians (2008) defines their scope of practice as follows: "The family physician's care is both personal and comprehensive and not limited by age, sex, organ system or type of problem, be it biological, behavioral, or social."

Questions regarding the number and type of assessments and treatments with which a professional psychologist, in various kinds of general and specialized practice, should be (or can be) proficient have not received extensive research attention. The ability to address physical health issues in behavioral health care also needs more examination. As medicine increasingly appreciates the importance of behavior to physical health, perhaps professional psychology should also give more attention to the interaction of behavior with physical health and sociocultural factors. These types of questions are difficult to evaluate when using traditional theoretical orientations to structure and organize education and practice in the field. The biopsychosocial approach, on the other hand, is very well suited for addressing these questions. That other fields also use this approach makes the evaluation of these questions much easier as well.

Implications for Professional Practice

The implications for education and licensure noted above apply here as well. The reconceptualization of the traditional theoretical orientations primarily as therapies would be a significant change for many psychologists. Incorporating an epidemiological perspective on behavioral health would also greatly expand the focus of assessment and treatment for many psychologists. Not only would a greater number of frequent behavioral health concerns involving addictions, sexuality, and other issues receive greater attention, but more attention would be focused on level of functioning in important life roles in addition to the traditional focus on psychological distress and symptoms. Much more attention would also be given to physical health and sociocultural factors as well. As graduate training expands to cover the full range of biopsychosocial needs (e.g., physical health habits and behaviors, chronic medical problems, addictions, sexuality issues, relationships and family functioning, child maltreatment, educational and vocational effectiveness), psychologists will be better able to assess and treat a broader range of biopsychosocial issues and their interactions.

The biopsychosocial approach fits very well with the recent interest in integrated health care models for meeting behavioral and physical health care needs (APA Presidential Task Force on the Future of Psychology Practice, 2009; Goodheart, 2010). Integrated health care involves psychologists being part of the health care team within primary care clinics so that behavioral health treatment is coordinated with primary care medical services (Bray & Rogers, 1995). Providing psychological services in these settings requires far more communication and collaboration than what is typical in most traditional psychotherapy practices.

The biopsychosocial approach also easily accommodates specializations in professional psychology that have newly developed or are recently growing. Several of these areas emphasize biological factors such as the specializations in health psychology, neuropsychology, sports psychology, geropsychology, psychopharmacology, and integrated behavioral and primary medical care. Others, such as executive coaching and correctional and forensic psychology, emphasize the interaction of the person and his or her environment. A factor that distinguishes many of these specializations is their relatively limited reliance on psychotherapy. As a result, the traditional theoretical orientations are less useful for case conceptualization and practice in these areas. The biopsychosocial approach, on the other hand, is very appropriate and useful for conceptualizing practice in all of these areas.

The emphasis of the biopsychosocial approach on treatment effectiveness is also consistent with the growing importance of outcomes and accountability in health care. Outcomes assessment undoubtedly will be increasingly integrated into all types of health care due to the potential for improving the effectiveness of prevention, intervention, and maintenance health care services. Improving behavioral health treatment effectiveness, prevention strategies, patient satisfaction with services, cost effectiveness, and patient functioning across the biopsychosocial domains would also lead to greater demand for psychological services and a potentially much larger role for behavioral health care in overall health care and human services.

Implications for Research and Science

The science-based and health care orientation of the biopsychosocial approach advocated in this volume rests on science for informing psychological practice and evaluating the safety and effectiveness of behavioral health care interventions. The biopsychosocial approach values research aimed at all levels of natural organization that help describe the inextricably intertwined biopsychosocial nature of human psychology. It takes the view that human psychology simply cannot be understood without this kind of integrative approach.

As a result of this comprehensive, integrative perspective, the biopsychosocial approach helps reduce the tendencies toward "schoolism" and partisanship that characterized the pre-paradigmatic era of the field. Many psychologists have held strong allegiances to theoretical orientations, schools of thought, and methodological approaches, and there is substantial evidence that allegiance effects have been a significant problem in health care research. A tendency to find that one's preferred treatment is more effective than others has consistently been found (Robinson, Berman, & Neimeyer, 1990; Rothstein, Sutton, & Borenstein, 2005; Smith, Glass, & Miller, 1980; Wampold, 2001). Such allegiance effects impede the progress of research and can seriously undermine the public's trust in the effectiveness or even the safety of psychotherapy and medical treatments. Schoolism also impedes research progress when researchers primarily address others from within their particular theoretical camps, as opposed to addressing the larger research and practice community. A commitment to well-controlled research and to questions of significance for the whole psychological community will help overcome these tendencies. From the biopsychosocial perspective, research is valued that clarifies any of the psychological, sociocultural, or biological influences on development, functioning, or behavior change. And what is especially valuable is compelling research that advances our understanding of how best to address behavioral health needs and promote biopsychosocial functioning. Strengthening the common purpose of the field will help reduce the divides that have grown between clinicians and researchers, quantitative and qualitative researchers, and the various theoretical schools and camps within the field. In addition, tendencies to develop new theoretical orientations or psychological interventions not based on validated scientific findings are also greatly reduced.

The health care orientation of the biopsychosocial approach will also focus more research on the effectiveness of clinical services and improving clinical practice. As noted earlier, many professional psychologists in the past appeared to view the field in terms of a service industry where psychologists could offer different types of psychological services to the public and patients could select and purchase services based on their preferences and needs. A stronger health care orientation to the field, however, focuses attention on meeting the behavioral health and biopsychosocial needs of the public. This in turn increases the importance of evaluating and improving clinical effectiveness. Applied health care fields often take a clinical science approach to research where the focus is on improving the effectiveness of clinical intervention through the use of experimental methods and applying the findings of laboratory research in clinical practice. For example, the Association of Clinical Scientists was formed in 1949 by a group of physicians and scientists to promote improvement in medical diagnosis, prognosis, and monitoring, and the application of scientific methods in clinical practice and research (Association of Clinical Scientists, 2010). A clinical science emphasis in professional psychology would likewise focus attention on the use of experimental research methods to improve clinical practice. This perspective is also consistent with the current emphasis of the National Institutes of Health on translating scientific findings into practical and useful interventions for clinical practice (Collins, 2010).

A development that will help propel the field in this direction is the improved methodology surrounding mental health outcomes research. Use of these new methods will undoubtedly lead to an improved understanding of the treatment process. Increased reliability and validity of outcome measurement, improved therapist and treatment data, and improved hardware and software computing capabilities can be used to identify the mechanisms of change and necessary conditions of effective psychological treatment for different types of patients, conditions, and therapists. Recently developed data mining statistical procedures can be used with very large data sets from actual practice settings to examine treatment and patient change in more detail than was ever possible before. Increasing the effectiveness of treatment, patient satisfaction with services, and cost effectiveness obviously would be extremely beneficial to the profession and the public, and would increase demand for psychological services as well. Identifying the mechanisms of change that account for the effectiveness of treatment may also help clarify the processes that originally led to psychopathology, maladjustment, and resilience. A more complete understanding of these issues would have implications for health care policy, prevention efforts, and social policy as well.

Conclusions

Psychology has undergone dramatic growth and development over its relatively short history. Though this growth has been tumultuous at times, the field has emerged as an influential scholarly discipline and clinical profession. It is also now reaching a particularly exciting time in its development. The science of psychology is progressing rapidly, with important and useful advances being made in many areas. The potential for professional psychology to expand its role in health care is also growing as medicine increasingly recognizes the importance of behavior to health and the United States is reconsidering its approach to health and health care.

It is particularly exciting for the field to reach the point where it is ready to leave behind its pre-paradigmatic past. The science of psychology has strengthened significantly in recent years, and it appears that a comprehensive biopsychosocial metatheoretical approach can now replace the traditional assortment of theoretical orientations that have been used for conceptualizing professional practice. Combining this scientific perspective with a grounding in professional ethics results in a very solid foundation for education, research, and practice in the field. When the field adopts this type of unified framework, the justification for many of the historical conflicts and divisiveness will weaken or even disappear. In their place will be a more unified sense of purpose and direction, and time and energy will be focused more efficiently on improving our understanding of human psychology, behavioral health care services, and the behavioral health and biopsychosocial functioning of the public in general.

Orienting the field around the behavioral health and biopsychosocial needs of the public expands the traditional conceptualization of the field in several important ways. This orientation increases the profession's focus on and responsibility for the health and functioning of individuals, families, communities, and society in general. Increasing collaboration with related professions and integrating behavioral health care into primary health care are logical outcomes of this broadened perspective. A comprehensive biopsychosocial orientation to professional psychology education may even be necessary before psychologists are well prepared to easily integrate into primary health care settings. Another logical outcome of this perspective is an increased emphasis on prevention, health promotion, and a public health perspective on behavioral health, because there are many issues where prevention is clearly the most effective, efficient, and humane approach to dealing with problems, as opposed to providing treatment after problems have already developed. Another important outcome of this approach will be increased attention to demonstrating the effectiveness of psychological services, both in terms of diagnostic and demographic groups who receive particular treatments and in terms of assessing outcomes in individual clinical cases.

The role of biomedical ethics also receives significant attention in the biopsychosocial approach advocated in this volume. Ethics have certainly always been important in behavioral health care, but the framework here gives professional ethics a central role in undergirding clinical practice. Because science cannot answer many questions routinely encountered in clinical practice, professional ethics are fundamental to the foundations on which professional practice lies. As a result, ethics training in professional psychology needs to be sufficiently broad and deep to ensure a solid foundation for balancing the interacting ethical, legal, and clinical issues that are regularly encountered in clinical practice. These issues are growing in importance as science, technology, multiculturalism, and globalism present new challenges and opportunities in health care and in society in general.

Professional psychology graduate and continuing education both will need to be nimble to stay current with these developments. The science of psychology is obviously progressing very quickly. If it is accurate that scientific knowledge is increasing at an exponential rate, the challenge of staying current is truly daunting. Biomedical ethics have also evolved quickly in recent decades and are a growing influence on health care research and practice. The challenge of applying ethical theory and principles across cultures, religions, and political perspectives has also proven to be complex. Attention needs to be placed on ensuring that students, faculty, and practitioners are staying current with all these developments.

If professional psychology makes progress in improving the behavioral health and biopsychosocial functioning of the general public, the potential for the profession to grow and extend its influence on health care is great. The prevalence and consequences of mental health and substance abuse disorders are substantial, and a relatively small proportion of the population is functioning optimally. The prevalence of chronic medical disease is very high, and educational, vocational, family, and social problems are pervasive and serious. Reducing the incidence and severity of these problems and improving individuals' health and functioning will be invaluable for individuals, families, communities, and society as a whole.

In some ways, professional psychology is uniquely situated to work toward improving individuals' health and well-being. Perhaps more than any other discipline, psychology possesses expertise for improving biopsychosocial functioning holistically. Psychologists are expert at dealing with the highly complex intrapersonal and interpersonal interactions between the biopsychosocial domains and at realizing the synergy that results when interventions are strategically aimed at resolving problems and building strengths across the biopsychosocial areas.

Endorsing the biopsychosocial approach will allow the field of professional psychology to resolve and leave behind controversies and conflicts that distracted it in the past and direct its energies with a unified perspective and sense of purpose toward improving behavioral health and biopsychosocial functioning. It will also put the field in a position where it can lead the effort to improve the health and functioning of individuals, families, and the general public. The field is now ready to leave behind its pre-paradigmatic past. And all the signs suggest that the paradigmatic era of professional psychology will be a very productive and exciting period for the profession.

References

Abramowitz, J. S. (1996). Variants of exposure and response prevention in the treatment of obsessive–compulsive disorder: A meta-analysis. *Behavior Therapy*, 27, 385–600.

Accreditation Council for Graduate Medical Education. (2011). *Common program requirements*. Retrieved from http://www.acgme.org/acWebsite/dutyHours/dh_dutyhours CommonPR07012007.pdf

Achenbach, T. M., & Edelbrock, C. S. (1983). *Manual for the child behavior checklist and revised child behavior profile*. Burlington, VT: Department of Psychiatry, University of Vermont.

Ackerman, S., & Hilsenroth, M. (2001). A review of therapist characteristics and techniques negatively impacting the therapeutic alliance. *Psychotherapy*, 38, 171–185.

Agras, W. S., Crow, S. J., Halami, K. A., Mitchell, J. E., Wilson, G. T., & Kraemer, H. C. (2000). A multicenter comparison of cognitive-behavioral therapy and interpersonal psychotherapy for bulimia nervosa. *American Journal of Psychiatry*, 157, 1302–1308.

Albee, G. W. (1959). *Mental health manpower trends*. New York: Basic Books.

Albee, G. W. (2006). Historical overview of primary prevention of psychopathology: Address to the third World Conference on the Promotion of Mental Health and Prevention of Mental and Behavioral Disorders, September 15–17, 2004, Auckland, New Zealand. *The Journal of Primary Prevention*, 27, 449–456.

American Academy of Family Physicians. (2008). *Family medicine, scope and philosophical statement*. Retrieved from http://www.aafp.org/online/en/home/policy/policies/f/scopephil.html

American Board of Professional Psychology. (2011a). *Specialty certification in clinical psychology*. Retrieved from http://www.abpp.org/files/page-specific/3355%20Clinical/01_Brochure.pdf

American Board of Professional Psychology. (2011b). *Specialty certification in counseling psychology*. Retrieved from http://www.abpp.org/files/page-specific/3364%20Counseling/01_Brochure.pdf

American Counseling Association. (2005). *Code of ethics and standards of practice*. Retrieved from http://www.counseling.org/Resources/CodeOfEthics/TP/Home/CT2.aspx

American Psychiatric Association. (1980). *Diagnostic and statistical manual for mental and emotional disorders* (3rd ed.). Washington, DC: Author.

American Psychiatric Association. (2000a). *Diagnostic and statistical manual for mental and emotional disorders* (4th ed.). Washington, DC: Author.

American Psychiatric Association. (2000b). *Therapies focused on attempts to change sexual orientation (reparative or conversion therapies: Position statement*. Retrieved from http://www.psych.org/Departments/EDU/Library/APAOfficialDocumentsandRelated/PositionStatements/200001.aspx

American Psychiatric Association. (2006). *Practice guidelines for the treatment of psychiatric disorders: Compendium 2006*. Arlington, VA: American Psychiatric Association.

American Psychological Association. (1996). *Recommended postdoctoral training in psychopharmacology for prescriptive privileges*. Retrieved from www.apa.org/ed/rx_pmodcurri.pdf

American Psychological Association. (2002). Ethical principles of psychologist and code of conduct. *American Psychologist*, 57, 1060–1075.

American Psychological Association. (2003). Guidelines in multicultural education, training, research, practice, and organizational change for psychologists. *American Psychologist*, 58, 377–402.

American Psychological Association. (2006). *Health care for the whole person statement of vision and principles*. Retrieved from www.apa.org/practice/hcwp_statement.html

American Psychological Association. (2007). Guidelines for psychological practice with girls and women. *American Psychologist*, 62, 949–979.

American Psychological Association. (2009). *Organizational purposes*. Retrieved from http://www.apa.org/about/

American Psychological Association. (2010). *APA mission statement*. Retrieved from http://www.apa.org/about/

American Psychological Association Assessment of Competency Benchmarks Work Group. (2007). *A developmental model for the defining and measuring competence in professional psychology*. Washington, DC: Author.

American Psychological Association Commission on Accreditation. (2007). *Policy statements and implementation regulations*. Washington, DC: American Psychological Association.

American Psychological Association Commission on Accreditation. (2009). *Guidelines and principles for accreditation of programs in professional psychology*. Washington, DC: American Psychological Association.

American Psychological Association Commission for the Recognition of Specialties and Proficiencies in Professional Psychology. (2011a). *Public description of clinical psychology*. Retrieved from http://www.apa.org/ed/graduate/specialize/clinical.aspx

American Psychological Association Commission for the Recognition of Specialties and Proficiencies in Professional Psychology. (2011b). *Public description of counseling psychology*. Retrieved from http://www.apa.org/ed/graduate/specialize/counseling.aspx

American Psychological Association Presidential Task Force on Evidence-Based Practice. (2006). Evidence-based practice in psychology. *American Psychologist*, 61, 271–285.

American Psychological Association Presidential Task Force on the Future of Psychology Practice. (2009). *Final report*. Washington, DC: American Psychological Association.

Anchin, J. C. (2008). Pursuing a unifying paradigm for psychotherapy: Tasks, dialectical considerations, and biopsychosocial systems metatheory. *Journal of Psychotherapy Integration*, 18, 310–349.

Anderson, R. L., & Lyons, J. S. (2001). Needs-based planning for persons with serious mental illness residing in intermediate care facilities. *Journal of Behavioral Health Services and Research*, 28, 104–110.

Annon, J. S. (1976). *Behavioral treatment of sexual problems*. Hagerstown, MD: Harper & Row.

Ariely, Dan (2009). *Predictably irrational: The hidden forces that shape our decisions*. New York: HarperCollins.

Association of Clinical Scientists. (2010). *Goals*. Retrieved from www.clinicalscience.org/info.html#goals

Association of Psychology Postdoctoral and Internship Centers. (2007). *Application for psychology internships*. Retrieved from http://www.appic.org/match/5_3_match_application.html

Association for Psychology Postdoctoral and Internship Centers. (2009). *APPIC application for psychology internship*. Retrieved from http://appic.org/match/5_3_match_application.html#PREVIOUSAAPI

Association of State and Provincial Psychology Boards (1998). *ASPPB model act for licensure of psychologists.* Retrieved from http://www.asppb.org/publications/model/act. aspx

Association for State and Provincial Psychology Boards. (2011). *Obtaining a license.* Retrieved from www.asppb.net/i4a/pages/index.cfm?pageid=3391.

Atherly, A. (2006). In R. L. Kane (Ed.), *Understanding health care outcomes research* (2nd ed., pp. 123–164). Sudbury, MA: Jones and Barlett.

Bader, C., Dougherty, K., Froese, P., Johnson, B., Menchen, F. C., & Park, J. Z., et al. *The Baylor religion survey.* Waco, TX: Baylor Institute for Religious Studies.

Bak, P. (1996). *How nature works: The science of self-organized criticality.* New York: Springer.

Baofu, P. (2007). *The future of complexity: Conceiving a better way to understand order and chaos.* Singapore: World Scientific Publishing.

Barkham, M., Margison, F., Leach, C., Lucock, M., Mellor-Clark, J., & Evans, C., et al. (2001). Service profiling and outcomes benchmarking using the CORE-OM: Toward practice-based evidence in the psychological therapies. *Journal of Consulting and Clinical Psychology, 69,* 184–196.

Barlow, D. H. (2004). Psychological treatments. *American Psychologist, 59,* 869–878.

Barlow, D. H. (2010). Negative effects from psychological treatments: A perspective. *American Psychologist, 65,* 13–20.

Barlow, D. H., Hayes, S. C., & Nelson, R. O. (1984). *The scientist-practitioner: Research and accountability in clinical and educational settings.* New York: Pergamon.

Barth, R. P., Scarborough, A. A., Lloyd, E. C., Losby, J. L., Casanueva, C., & Mann, T. (2008). *Developmental status and early intervention services needs of maltreated children: Final report.* Washington, DC: U.S. Department of Health and Human Services, Office of the Assistant Secretary for Planning and Evaluation.

Bateman, A., & Fonagy, P. (2008). 8-Year follow-up of patients treated for borderline personality disorder: Mentalization-based treatment versus treatment as usual. *The American Journal of Psychiatry, 165,* 631–638.

Beauchamp, T. L., & Childress, J. F. (1977). *Principles of biomedical ethics.* New York: Oxford University Press.

Beauchamp, T. L., & Childress, J. F. (2009). *Principles of biomedical ethics* (6th ed.). New York: Oxford University Press.

Beauchamp, T. L., Walters, L., Kahn, J. P., & Mastroianni, A. C. (2008). *Contemporary issues in bioethics.* Belmont, CA: Thomson Wadsworth.

Beck, A. T., Rush, J. A., Shaw, B. F., & Emery, G. (1979). *Cognitive therapy for depression.* New York: Guilford.

Beck, A. T., Steer, R. A., & Brown, G. K. (1996). *Manual for the Beck Depression Inventory-II.* San Antonio, TX: Psychological Corporation.

Benjamin, L. T., Jr. (2007). *A brief history of modern psychology.* Malden, MA: Blackwell.

Bergin, A. E. (1966). Some implications of psychotherapy: Negative results revisited. *Journal of Counseling Psychology, 10,* 224–250.

Bergin, A. E. (1971). The evaluation of therapeutic outcomes. In A. E. Bergin, & S. L. Garfield (Eds.), *Handbook of psychotherapy and behavior change* (pp. 217–270). New York: John Wiley and Sons.

Beutler, L. E., Malik, M., Alimohamed, S., Harwood, T. M., Talebi, H., & Noble, S., et al. (2004). Therapist variables. In M. J. Lambert (Ed.), *Garfield and Bergin's handbook of psychotherapy and behavior change: An empirical analysis* (5th ed., pp. 227–306). New York: Wiley.

Beutler, L. E., Malik, M., Talebi, H., Fleming, J., & Moleiro, C. (2004). Use of psychologi-cal tests/instruments for treatment planning. In M. E. Maruish (Ed.), *The use of psychological testing for treatment planning and outcomes assessment* (3rd ed., pp. 11–146). Mahwah, NJ: Lawrence Erlbaum.

Beutler, L. L. (1983). *Eclectic psychotherapy: A systematic approach*. Elmsford, NY: Pergamon.

Binggeli, N. J., Hart, S. N., & Brassard, M. R. (2001). *Psychological maltreatment: A study guide*. Thousand Oaks, CA: Sage.

Blaisure, K. R., & Geasler, M. J. (2006). Educational interventions for separating and divorc-ing parents and their children. In M. A. Fine, & J. H. Harvey (Eds.), *Handbook of divorce and relationship dissolution* (pp. 575–602). Mahwah, NJ: Lawrence Erlbaum.

Bloom, M., & Fischer, J. (1982). *Evaluating practice: Guidelines for the accountable professional*. Englewood Cliffs, NJ: Prentice-Hall.

Bloom, M., Fischer, J., & Orme, J. G. (2003). *Evaluating practice: Guidelines for the accountable professional* (4th ed.). Boston: Allyn & Bacon.

Blount, A., Scheonbaum, M., Kathol, R., Rollman, B., Thomas, M., & O'Donohue, W., et al. (2007). The economics of behavioral health services in medical settings: A summary of the evidence. *Professional Psychology: Research and Practice, 38*, 290–297.

Boahen, K. (2005). Neuromorphic microchips. *Scientific American, 292*, 56–63.

Bonnano, G. A., & Lilienfeld, S. O. (2008). Let's be realistic: When grief counseling is effective and when it's not. *Professional Psychology: Research and Practice, 39*, 377–380.

Bosch, M., Faber, M. J., Cruigsberg, J., Voerman, G. E., Leatherman, S., & Grol, R. P. T. M., et al. (2010). Effectiveness of patient care teams and the role of clinical expertise and coordination: A literature review. *Medical Care Research and Review, 66*, 5S–35S. doi:10.1177/1077558709343295.

Bray, J. H., & Rogers, J. C. (1995). Linking psychologists and family physicians for collabo-rative practice. *Professional Psychology: Research and Practice, 26*, 132–138.

Bringhurst, D. L., Watson, C. W., Miller, S. D., & Duncan, B. L. (2006). The reliability and validity of the Outcome Rating Scale: A replication study of a brief clinical measure. *Journal of Brief Therapy, 5*, 23–30.

Britner, P. A. (2007). Editorial—Reflections on the life and work of George Albee: Introduction to the Special Issue. *The Journal of Primary Prevention, 28*, 1–2. doi:10.1007/s10935-006-0079-z.

Bronfenbrenner, U. (1979). *The ecology of human development*. Cambridge, MA: Harvard University Press.

Buss, D. M. (1991). Evolutionary personality psychology. *Annual Review of Psychology, 42*, 459–491.

Buss, D. M. (1995). Evolutionary psychology: A new paradigm for psychological science. *Psychological Inquiry, 6*, 1–49.

Byrne, D. (1998). *Complexity theory and the social sciences: An introduction*. London: Routledge.

Califano, J. A. (1979). The Secretary's forward. In J. B. Richmond (Ed.), *Healthy people: The Surgeon General's report on health promotion and disease prevention* (pp. vii–x). Washington, DC: Government Printing Office.

Callahan, D. (1990). *What kind of life*. New York: Simon & Schuster.

Campbell, W. H., & Rohrbaugh, R. M. (2006). *The biopsychosocial formulation manual: A guide for mental health professionals*. New York: Routledge Taylor & Francis Group.

Cannon, W. B. (1932). *The wisdom of the body*. New York: Norton.

Capra, F. (1975). *The Tao of physics*. Boston: Shambala.

Capra, F. (1996). *The web of life: A new scientific understanding of living systems*. New York: Anchor Books.

Capra, F. (2002). *The hidden connections: A science for sustainable living*. New York: Anchor Books.

Carr, A., & McNulty, M. (2006). *The handbook of adult clinical psychology: An evidence-based practice approach*. East Sussex, England: Routledge.

Cassidy, J., & Shaver, P. R. (Eds.), (2008). *Handbook of attachment: Theory, research and clinical applications* (2nd ed.). New York: Guilford.

Center for Disease Control. (2005). *Summary health statistics for U.S. adults: National health interview survey, 2005*. Retrieved at http://www.cdc.gov/nchs/data/series/sr_10/sr10_232.pdf

Center for Disease Control. (2007). Deaths: Final data for 2004. *National Vital Statistics Reports*, **55** (19), 1–120. Retrieved from http://www.cdc.gov/nchs/data/nvsr/nvsr55/nvsr55_19.pdf

Center for Disease Control. (2009). Cigarette smoking among adults and trends in smoking cessation—United States, 2008. *MMWR Weekly*. Retrieved from www.cdc.gov/mmwr/preview/mmwrhtml/mm5844a2.htm

Centers for Disease Control and Prevention and National Institute of Justice. (2000). *Extent, nature, and consequences of intimate partner violence*. Rockville, MD: National Criminal Justice Reference Service.

Centerwall, S. A., & Centerwall, W. R. (2000). The discovery of phenylketonuria: The story of a young couple, two affected children, and a scientist. *Pediatrics*, 105, 89–103.

Chaffin, M., Hanson, R., Saunders, B. E., Nichols, T., Barnett, D., & Zeanah, C., et al. (2006). Report of the APSAC Task Force on attachment therapy, reactive attachment disorder, and attachment problems. *Child Maltreatment*, 11, 76–89. doi:10.1177/1077559505283699.

Chambless, D. L., & Hollon, S. D. (1998). Defining empirically supported therapies. *Journal of Consulting and Clinical Psychology*, 66, 7–18.

Chiles, A., & Strosahl, D. (2005). *Clinical manual for assessment and treatment of suicidal patients*. Washington, DC: American Psychiatric Publishing.

Clark, D. M., Layard, R., Smithies, R., Richards, D. A., Suckling, R., & Wright, B. (2009). Improving access to psychological therapy: Initial evaluation of two UK demonstration sites. *Behavior Research and Therapy*, 47, 910–920.

Clay, R. A. (2009). Screening, brief intervention, and referral to treatment: New populations, new effectiveness data. *SAMHSA News*, *17*(6), 1–3.

Clement, P. W. (1999). *Outcomes and incomes: How to evaluate, improve, and market your psychotherapy practice by measuring outcomes*. New York: Guilford.

Cohen, D. (2004). Boston and the history of biomagnetism. *Neurology and Clinical Neurophysiology* 30.

Cohen, E. D., & Cohen, G. S. (1999). *The virtuous therapist: Ethical practice of counseling and psychotherapy*. Belmont, CA: Brooks/Cole Wadsworth.

Cohen, J. (1988). *Statistical power analysis for the behavioral sciences* (2nd ed). Hillsdale, NJ: Lawrence Erlbaum Associates.

Collins, F. S. (2010). Opportunities for Research and NIH. *Science*, 327, 36–37.

Collins, F. S., Green, E. D., Guttmacher, A. E., & Guyer, M. S. (2003). A vision for the future of genomics research: A blueprint for the genomic era. *Nature*, 422, 835–847.

Commission on Chronic Illness. (1957). *Chronic illness in the United States* (Vol. 1). Cambridge, MA: Harvard University Press.

Confer, J. C., Easton, J. A., Fleischman, D. S., Goetz, C. D., Lewis, D. M. G., & Perilloux, C., et al. (2010). Evolutionary psychology: Controversies, questions, prospects, and limitations. *American Psychologist, 65,* 110–126.

Corey, G., Corey, M. S., & Callahan, P. (2003). *Issues and ethics in the helping professions* (6th ed.). Pacific Grove, CA: Brooks/Cole.

Corsini, R. J., & Wedding, D. (2008). *Current psychotherapies* (8th ed.). Belmont, CA: Thomson Brooks/Cole.

Council for the Parliament of the World's Religions (1993). *Declaration toward a global ethic.* Retrieved from http://www.parliamentofreligions.org/index.cfm?n = 4&sn = 4

Cowen, E. L. (1994). The enhancement of psychological wellness: Challenges and opportunities. *American Journal of Community Psychology, 22,* 149–179.

Crump, T. (2001). *A brief history of science: As seen through the development of scientific instruments.* London: Constable.

Crutchfield, J. P., Farmer, J. D., Packard, N. H., & Shaw, R. S. (1986). Chaos. *Scientific American,* Dec, 255.

Cummings, N. A. (2005). Resolving the dilemmas in mental healthcare delivery: Access, stigma, fragmentation, conflicting research, politics and more. In N. A. Cummings, W. T. O'Donohue, & M. A. Cucciare (Eds.), *Universal healthcare: Readings for mental health professionals* (pp. 47–74). Reno, NV: Context Press.

Cummings, N. A., & O'Donohue, W. T. (2008). *Eleven blunders that cripple psychotherapy in America: A remedial unblundering.* New York: Taylor & Francis.

Cummings, N. A., O'Donohue, W. T., & Cucciare, M. A. (Eds.), (2005). *Universal healthcare: Readings for mental health professionals* Reno, NV: Context Press.

Dar, A. (2006). The new astronomy. In G. Fraser (Ed.), *The new physics for the 21st century* (pp. 69–85). Cambridge, UK: Cambridge University Press.

Davidson, G. C. (2000). Stepped care: Doing more with less? *Journal of Consulting and Clinical Psychology, 68,* 580–585.

Davidson, M. (1983). *Uncommon sense: The life and thought of Ludwig von Bertalanffy.* Los Angeles: J. P. Tarcher.

Davis, M. K., & Gidycz, C. A. (2000). Child sexual abuse prevention programs: A meta-analysis. *Journal of Clinical Child Psychology, 29,* 257–265.

Davis, K., Schoen, C., & Stremikis, K. (2010). *Mirror, mirror on the wall: How the performance of the U.S. health care system compares internationally.* New York: The Commonwealth Fund.

Dawkins, R. (1976). *The selfish gene.* New York: Oxford University Press.

Derogatis, L. R., & Melisaratos, N. (1983). The Brief Symptom Inventory: An introductory report. *Psychological Medicine, 13,* 595–605.

Dollard, J., & Miller, N. E. (1950). *Personality and psychotherapy.* New York: McGraw-Hill.

Donagan, A. (1977). *The theory of morality.* Chicago: University of Chicago Press.

Driver-Linn, E. (2003). Where is psychology going? Structural fault lines revealed by psychologists' use of Kuhn. *American Psychologist, 58,* 269–278.

Drum, D. J., Brownson, C., Denmark, A. B., & Smith, S. E. (2009). New data on the nature of suicidal crises in college students: Shifting the paradigm. *Professional Psychology: Research and Practice, 40,* 213–222.

Dubos, R. (1959). *Mirage of health.* New York: Harper & Row.

Duffy, F. F., Wilk, J., West, J. C., Narrow, W. E., Rae, D. S., & Hall, R., et al. (2006). Chapter 21: Mental health practitioners and trainees. In R. W. Manderscheid, & J. T. Berry (Eds.), *Mental Health: United States, 2004.* Retrieved from http://www. mentalhealth.samhsa.gov/publications/allpubs/SMA06-4195/Chapter22.asp

Durbin, J., Goering, P., Cochrane, J., Macfarlane, D., & Sheldon, T. (2004). Needs-based planning for persons with schizophrenia residing in board-and-care homes. *Schizophrenia Bulletin, 30,* 123−132.

Durlak, J. A., & Wells, A. M. (1997). Primary prevention mental health programs for children and adolescents: A meta-analytic review. *American Journal of Community Psychology, 25,* 115−152.

Dworkin, R. (1977). *Taking rights seriously.* Cambridge, MA: Harvard University Press.

Dziegielewski, S. F. (2010). *DSM-IV-TR in action* (2nd ed.). Hoboken, NJ: John Wiley & Sons.

Eddy, D. (1992). Cost-effectiveness analysis: Is it up to the task? *Journal of the American Medical Association, 267,* 3344.

Eells, T. D. (2007). *Handbook of psychotherapy case formulation* (2nd ed.). New York: Guilford.

Egger, H. L., & Emde, R. N. (2011). Developmentally sensitive diagnostic criteria for mental health disorders in early childhood: The Diagnostic and Statistical Manual of Mental Disorders-IV, the Research Diagnostic Criteria—Preschool Age, and the Diagnostic Classification of Mental Health and Developmental Disorders in Infancy and Early Childhood-Revised. *American Psychologist, 66,* 95−106.

Elkin, I. (1994). The NIMH treatment of depression collaborative research program. Where we began and where we are. In A. E. Bergin, & S. L. Garfield (Eds.), *Handbook of psychotherapy and behavior change* (4th ed., pp. 114−142). New York: Wiley.

Ellenbogen, M. A., Ostiguy, C. S., & Hodgins, S. (2010). Intergenerational effects of high neuroticism in parents and their public health significance. *American Psychologist, 65,* 135−136.

Ellis, A. (1973). *Humanistic psychotherapy: The rational-emotive approach.* New York: McGraw-Hill.

Emmott, S., & Rison, S. (2006). *Towards 2020 science.* Cambridge, UK: Microsoft Research.

Encyclopædia Britannica. (2010). *Metatheory.* Retrieved from http://www.britannica.com/EBchecked/topic/378037/metatheory

Engel, G. L. (1977). The need for a new medical mode: A challenge for biomedicine. *Science, 196,* 129−136.

Eysenck, H. J. (1952). The effects of psychotherapy: An evaluation. *Journal of Consulting Psychology, 16,* 319−324.

Eysenck, H. J. (1997). Personality and experimental psychology: The unification of psychology and the possibility of a paradigm. *Journal of Personality and Social Psychology, 73,* 1224−1237.

Feigenbaum, M. J. (1980). Universal behavior in nonlinear systems. *Los Alamos Science, 1,* 4−27.

Fennel, M. J. V., & Teasdale, J. D. (1987). Cognitive therapy for depression: Individual differences and the process of change. *Cognitive Therapy and Research, 11,* 253−271.

Finkelhor, D. (1994). Current information on the scope and nature of child sexual abuse. *Future of Children, 4,* 31−53.

Finkelhor, D., & Dziuba-Leatherman, J. (1994). Children as victims of violence: A national survey. *Pediatrics, 94,* 413−420.

Flexner, A. (1910). *Medical education in the United States and Canada: A report to the Carnegie Foundation for the Advancement of Teaching, Bulletin No. 4.* New York: Carnegie Foundation for the Advancement of Teaching.

Flora, J. A., Maccoby, N., & Farquhar, J. W. (1989). Communication campaigns to prevent cardiovascular disease: The Stanford Community Studies. In R. Rice, & C. Atkin (Eds.), *Public communication campaigns* (pp. 233−252). Thousand Oaks, CA: Sage.

Fournier, J. C., DeRubeis, R. J., Hollon, S. D., Dimidgian, S., Amsterdam, J. D., & Shelton, R. C., et al. (2010). Antidepressant drug effects and depression severity: A patient-level meta-analysis. *Journal of the American Medical Association, 303,* 47−53.

Fox, R. E., DeLeon, P. H., Newman, R., Sammons, M. R., Dunivin, D. L., & Baker, D. C. (2009). Prescriptive authority and psychology: A status report. *American Psychologist, 64,* 257−268.

Frank, J. D. (1961). *Persuasion and healing.* Baltimore: Johns Hopkins.

Frankel, R. M., Quill, T. E., & McDaniel, S. H. (Eds.), (2003). *The biopsychosocial approach: Past, present, and future* Rochester, NY: University of Rochester Press.

Freeman, S. J. (2000). *Ethics: An introduction to philosophy and practice.* Belmont, CA: Wadsworth.

French, T. M. (1933). Interrelations between psychoanalysis and the experimental work of Pavlov. *American Journal of Psychiatry, 89,* 1165−1203.

Fries, J. F., Koop, C. E., Beadle, C. E., Cooper, P. P., England, M. J., & Greaves, R. F., et al. (1993). Reducing health care costs by reducing the need and demand for medical services. *New England Journal of Medicine, 329,* 321−325.

Froyd, J., Lambert, M. J., & Froyd, J. (1996). A review of practices of psychotherapy outcome measurement. *Journal of Mental Health (UK), 5,* 11−15.

Garb, H. N. (1998). *Studying the clinician: Judgment research and psychological assessment.* Washington, DC: American Psychological Association.

Gardner, H. (2005). Scientific psychology: Should we bury it or praise it? In R. J. Sternberg (Ed.), *Unity in psychology: Possibility or pipedream?* (pp. 77−90). Washington, DC: American Psychological Association.

Garfield, S. L. (1994). Research on client variables in psychotherapy. In A. E. Bergin, & S. L. Garfield (Eds.), *Handbook of psychotherapy and behavior change* (4th ed., pp. 190−228). New York: Wiley.

Garmezy, N., & Rutter, M. (Eds.), (1983). *Stress, coping, and development in children* Baltimore, MD: Johns Hopkins University Press.

Gay, P. (1988). *Freud: A life for our times.* New York: Norton.

Gert, B., Culver, C. M., & Clouser, K. D. (1997). *Bioethics: A return to fundamentals.* New York: Oxford University Press.

Gert, B., Culver, C. M., & Clouser, K. D. (2006). *Bioethics: A systematic approach* (2nd ed.). New York: Oxford University Press.

Gilligan, C. (1982). *In a different voice: Psychological theory and women's development.* Cambridge, MA: Harvard University Press.

Goodheart, C. D. (2010). Economics and psychology practice: What we need to know and why. *Professional Psychology: Research and Practice, 41,* 89−195.

Goodheart, C. D., & Carter, J. A. (2008). The proper focus of evidence-based practice in psychology: Integration of possibility and probability. In W. B. Walsh (Ed.), *Biennial Review of Counseling Psychology* (Vol. 1, pp. 47−70). New York: Taylor & Francis Group.

Goodheart, C. D., & Kazdin, A. E. (2006). Introduction. In C. D. Goodheart, A. E. Kazdin, & R. J. Sternberg (Eds.), *Evidence-based psychotherapy: Where practice and research meet* (pp. 3−10). Washington, DC: American Psychological Association.

Gordon, D. M. (2007). Control without hierarchy. *Nature, 446(7132),* 143.

Gottman, J., Swanson, C., & Swanson, K. (2002). A general systems theory of marriage: Nonlinear difference equation modeling of marital interaction. *Personality and Social Psychology Review, 6,* 326−340.

Gottman, J. M. (1994). *What predicts divorce: The measures.* Hillsdale, NJ: Lawrence Erlbaum.

Gottman, J. M. (1999). *The marriage clinic*. New York: Norton.

Gould, R. A., Otto, M. W., & Pollack, M. H. (1995). A meta-analysis of treatment outcome for panic disorder. *Clinical Psychology Review*, 15, 819–844.

Gould, R. A., Otto, M. W., Pollack, M. H., & Yap, L. (1997). Cognitive behavioral and pharmacological treatment of generalized anxiety disorder: A preliminary meta-analysis. *Behavior Therapy*, 28, 285–305.

Greene, B. (1999). *The elegant universe: Superstrings, hidden dimensions, and the quest for the ultimate theory*. New York: Vintage books.

Greene, J. (2003). From neural "is" to moral "ought": What are the moral implications of neuroscientific moral psychology? *Nature Reviews: Neuroscience*, 4, 847–850.

Greene, J. D., Nystrom, L. E., Engell, A. D., Darley, J. M., & Cohen, J. D. (2004). The neural bases of cognitive conflict and control in moral judgment. *Neuron*, 44, 389–400.

Griffin, J. (1986). *Well-being: Its meaning, measurement and moral importance*. Oxford: Clarendon.

Grissom, R. J. (1996). The magical number 7 + − 2: Meta-meta-analysis of the probability of superior outcome in comparisons involving therapy, placebo, and control. *Journal of Consulting and Clinical Psychology*, 64, 973–982.

Grob, G. N. (1995). *The mad among us: A history of the care of America's mentally ill*. Cambridge, MA: Harvard University Press.

Gros, C. (2008). *Complex and adaptive dynamical systems: A primer*. Berlin: Springer-Verlag.

Groth-Marnat, G. (2009). *Handbook of psychological assessment* (5th ed.). Hoboken, NJ: John Wiley & Sons.

Guastello, S. J. (1995). *Chaos, catastrophe, and human affairs: Nonlinear dynamics in work, organizations, and social evolution*. Mahwah, MJ: Lawrence Erlbaum Associates.

Guastello, S. J. (2002). *Managing emergent phenomena: Nonlinear dynamics in work organizations*. Mahwah, MJ: Lawrence Erlbaum Associates.

Guastello, S. J. (2009). Group dynamics: Adaptation, coordination, and the emergence of leaders. In S. J. Guastello, M. Koopmans, & D. Pincus (Eds.), *Chaos and complexity in psychology: The theory of nonlinear dynamical systems* (pp. 402–433). New York: Cambridge University Press.

Guastello, S. J., Koopmans, M., & Pincus, D. (Eds.), (2009). *Chaos and complexity in psychology: The theory of nonlinear dynamical systems* New York: Cambridge University Press.

Guastello, S. J., & Liebovitch, L. S. (2009). Introduction to nonlinear dynanics and complexity. In S. J. Guastello, M. Koopmans, & D. Pincus (Eds.), *Chaos and complexity in psychology: The theory of nonlinear dynamical systems* (pp. 1–40). New York: Cambridge University Press.

Haas, E., Hill, R., Lambert, M. J., & Morrell, B. (2002). Do early responders to psychotherapy maintain treatment gains? *Journal of Clinical Psychology*, 58, 1157–1172.

Harris Interactive (2006). *Religious views and beliefs vary greatly by country*. Retrieved from www.harrisinteractive.com/news/printerfriend/index.asp?NewsID=1131

Harvey, D. L., & Reed, M. H. (1994). The evolution of dissipative social systems. *Journal of Social and Evolutionary Systems*, 17, 371–411.

Hauser, M. D. (2006). *Moral minds: How nature designed our universal sense of right and wrong*. New York: Ecco/HarperPerennial.

Hawking, S. (1996). *A brief history of time*. New York: Bantam Books.

Hawkins, G. S. (1983). *Mindsteps to the cosmos*. HarperCollins.

Hawkins, R. P., Mathews, J. R., & Hamdan, L. (1999). *Measuring behavioral health outcomes: A practical guide*. New York: Plenum.

Hayes, S. C., Strosahl, K. D., & Wilson, K. G. (1999). *Acceptance and commitment therapy: An experiential approach to behavior change.* New York: Guilford.

Hazelton. (2011). Recovery management. Retrieved from www.hazelton.org/web/public/continuingrecovery.page

Hazlett-Stevens, H., Craske, M. G., Roy-Byrne, P. P., Sherbourne, C. D., Stein, M. B., & Bystritsky, A. (2002). Predictors of willingness to consider medication and psychosocial treatment for panic disorder in primary care patients. *General Hospital Psychiatry, 24,* 316–321.

Henriques, G. R. (2003). The tree of knowledge system and the theoretical unification of psychology. *Review of General Psychology, 7,* 150–182.

Henriques, G. R. (2005). A new vision for the field: Introduction to the second special issue on the unified theory. *Journal of Clinical Psychology, 61,* 3–6.

Henriques, G. R., & Cobb, H. C. (2004). Introduction to the special issues on the unified theory. *Journal of Clinical Psychology, 60,* 1203–1205.

Henriques, G. R., & Sternberg, R. J. (2004). Unified professional psychology: Implications for the combined-integrated model of doctoral training. *Journal of Clinical Psychology, 60,* 1051–1063.

Hersen, M., & Rosqvist, J. (2008). *Handbook of psychological assessment, case conceptualization, and treatment.* Hoboken, NJ: John Wiley & Sons.

Hey, T., & Trefethen, A. (2003). The data deluge: An e-science perspective. In F. Berman, A. Hey, & G. Fox (Eds.), *Grid computing—making the global infrastructure a reality* (pp. 36–809–824). New York: John Wiley & Sons.

Hiatt, M. D., & Stockton, C. G. (2003). The impact of the Flexner Report on the fate of medical schools in North America after 1909. *Journal of American Physicians and Surgeons, 8,* 37–40.

Hickson, G. B., Clayton, E. W., Entman, S. S., Miller, C. S., Githens, P. B., & Whetten-Goldstein, K., et al. (1994). Obstetricians' prior malpractice experience and patients' satisfaction with care. *Journal of the American Medical Association, 272,* 1583–1587.

Hinshaw, S. P., & Stier, A. (2008). Stigma as related to mental disorders. *Annual Review of Clinical Psychology, 4,* 367–393.

Hirschfeld, R. M. A., & Russell, J. M. (1997). Assessment and treatment of suicidal patients. *New England Journal of Medicine, 333,* 910–915.

Hobbes, T. (1651/2002). *Leviathan, or the matter, forme, and power of a commonwealth, ecclesiasticall and civil.* Project Gutenberg.

Hofmann, S. G., Barlow, D. H., Papp, L. A., Detweiler, M. F., Ray, S. E., & Shear, M. K., et al. (1998). Pretreatment attrition in a comparative treatment outcome study on panic disorder. *American Journal of Psychiatry, 155,* 43–47.

Hollis, G., Kloos, H, & Van Orden, G. C. (2009). Origins of order in cognitive activity. In S. J. Guastello, M. Koopmans, & D. Pincus (Eds.), *Chaos and complexity in psychology: The theory of nonlinear dynamical systems* (pp. 206–241). New York: Cambridge University Press.

Hollon, S. D., & Beck, A. T. (2004). Cognitive and cognitive-behavioral therapies. In M. J . Lambert (Ed.), *Garfield and Bergin's handbook of psychotherapy and behavior change: An empirical analysis* (5th ed., pp. 447–492). New York: Wiley.

Holmes, O. W. (1939). In F. Frankfurter (Ed.), *Mr. Justice Holmes and the Supreme Court.* Cambridge, MA: Harvard University Press.

Horvath, A. O., & Bedi, R. P. (2002). The alliance. In J. C. Norcross (Ed.), *Psychotherapy relationships that work: Therapist contributions and responsiveness to patients* (pp. 37–70). New York: Oxford University Press.

Howard, K. I., Moras, K., Brill, P. L., Martinovich, Z., & Lutz, W. (1996). The evaluation of psychotherapy: Efficacy, effectiveness, and patient progress. *American Psychologist,* 51, 1059–1064.

Huyse, F. J., Lyons, J. S., Stiefel, F., Slaets, J., de Jonge, P., & Latour, C. (2001). Operationalizing the biopsychosocial model: The INTERMED. *Psychosomatics,* 42, 5–13.

Ilardi, H. H., & Craighead, W. E. (1994). The role of non-specific factors in cognitive-behavioral therapy for depression. *Clinical Psychology: Science and Practice,* 1, 138–156.

Institute of Medicine. (1994). *Reducing risks for mental disorders: Frontiers for preventive intervention research.* Washington, DC: National Academy Press.

Institute of Medicine. (2000). *To err is human: Building a safer health system.* Washington, DC: National Academy Press.

Institute of Medicine. (2001). *Crossing the quality chasm: A new health system for the 21st century.* Washington, DC: National Academy Press.

Institute of Medicine. (2004). *Improving medical education: Enhancing the behavioral and social science content of medical school curricula.* Washington, DC: Author.

International Union of Psychological Science (2005). *Universal declaration of ethical principles for psychologists.* Website.

Jane-Llopis, E., Hosman, C., Jenkins, R., & Anderson, P. (2003). Predictors of efficacy in depression prevention programmes. Meta-analysis. *British Journal of Psychiatry,* 183, 384–397.

Jecker, N. S., Jonsen, A. R., & Pearlman, R. A. (2007). *Bioethics: An introduction to the history, methods, and practice* (2nd ed.). Sudbury, MA: Jones & Bartlett.

Johnson, N. G. (2001). Psychology's mission includes health: an opportunity. *APA Monitor,* 32(4), 5.

Johnson, S. L. (2003). *Therapist's guide to clinical intervention: The 1-2-3's of treatment planning* (2nd ed.). San Diego: Academic Press.

Joiner, T. E. (2005). *Why people die by suicide.* Cambridge, MA: Harvard University Press.

Joint Commission on Accreditation of Healthcare Organizations. (2006). *Comprehensive accreditation manual for behavioral health care: 2006–2006, Standards, rationales, elements of performance, scoring.* Oakbrook Terrace, IL: Joint Commission Resources.

Jones, J. (1981). *Bad blood: The Tuskegee syphilis experiment.* New York: Free Press.

Jongsma, A. E., Berghuis, D. J., & Bruce, T. J. (2008). *The severe and persistent mental illness treatment planner.* Hoboken, NJ: John Wiley & Sons.

Jongsma, A. E., Peterson, L. M., & Bruce, T. J. (2006). *The complete adult psychotherapy treatment planner.* Hoboken, NJ: John Wiley & Sons.

Jongsma, A. E., Peterson, L. M., McInnis, W. P., & Bruce, T. J. (2006). *The child psychotherapy treatment planner.* Hoboken, NJ: John Wiley & Sons.

Judd, L. L. (1997). The clinical course of unipolar major depressive disorders. *Archives of General Psychiatry,* 54, 989–991.

Kandel, E. R. (2006). *In search of memory: The emergence of a new science of mind.* New York: W. W. Norton & Company.

Kane, R. L. (Ed.), (2006). *Understanding health care outcomes research.* (2nd ed.). Sudbury, MA: Jones & Barlett Publishers.

Kant, I. (1785/1964). *Groundwork of the metaphysics of morals.* Trans. H. J. Paton. New York: Harper & Row.

Kaslow, N. J. (2004). Competencies in professional psychology. *American Psychologist,* 59, 774–781.

Kaslow, N. J., Bollini, A. M., Druss, B., Glueckauf, R. L., Goldfrank, L. R., & Kelleher, K. J., et al. (2007). Health care for the whole person: Research update. *Professional Psychology: Research and Practice*, 38, 278–289.

Kaslow, N. J., Borden, K. A., Collins, F. L., Forrest, L., Illfelder-Kaye, J., & Nelson, P. D., et al. (2004). Competencies conference: Future directions in education and credentialing in professional psychology. *Journal of Clinical Psychology*, 60, 699–712.

Kaslow, N. J., Rubin, N. J., Bebeau, M. J., Leigh, I. W., Lichtenberg, J. W., & Nelson, P. D., et al. (2007). Guiding principles and recommendations for the assessment of competence. *Professional Psychology: Research and Practice*, 38, 441–451.

Kazak, A. E., Hoagwood, K., Weisz, J. R., Hood, K., Kratochwill, T. R., & Vargas, L. A., et al. (2010). A meta-systems approach to evidence-based practice with children and adolescents. *American Psychologist*, 65, 85–97.

Kelly, E. L. (1961). Clinical psychology—1960. Report of survey findings. *Newsletter: Division of Clinical Psychology of the American Psychological Association*, 14, 1–11.

Kendler, H. H. (2002). A personal encounter with psychology (1937–2002). *History of Psychology*, 5, 52–84.

Kessler, R. C., Demler, O., Frank, R. G., Olfson, M., Pincus, H. A., & Walters, E. E., et al. (2005). Prevalence and treatment of mental disorders, 1990 to 2003. *The New England Journal of Medicine*, 352, 2515–2524.

Kessler, R. C., Heeringa, S., Lakoma, M. D., Petukhova, M., Rupp, A. E., & Schoenbaum, M., et al. (2008). Individual and societal effects of mental disorders on earnings in the United States: Results from the National Comorbidity Survey Replication. *The American Journal of Psychiatry*, 165, 703–711.

Kessler, R. C., McGonagle, K. A., Zhao, S., Nelson, C. B., Hughes, M., & Eshleman, H. U., et al. (1994). Lifetime and 12-month prevalence of DSM-III-R disorders in the United States: Results from the National Comorbidity Study. *Archives of General Psychiatry*, 51, 8–19.

Keyes, C. L. M. (2007). Promoting and protecting mental health as flourishing: A complementary strategy for improving national mental health. *American Psychologist*, 62, 95–108.

Keynes, J. M. (1930/1972). *Economic possibilities for our grandchildren*. In *The Collected Writings of John Maynard Keynes*. London: Macmillan.

Kiresuk, T. J., & Sherman, R. E. (1968). Goal Attainment Scaling: A general method for evaluating comprehensive community mental health programs. *Community Mental Health Journal*, 4, 443–452.

Kitchner, K. (1984). Intuition, critical evaluation and ethical principles: The foundation for ethical decision in counseling psychology. *The Counseling Psychologist*, 12, 43–55.

Knapp, S., & VandeCreek, L. (2007). When values of different cultures conflict: Ethical decision making in a multicultural context. *Professional Psychology: Research and Practice*, 388, 660–666.

Kohn, L. T., Corrigan, J. M., & Donaldson, M. S. (Eds.), (2000). *To err is human: Building a safer health system* Washington, DC: National Academy Press.

Koocher, G. P., & Keith-Spiegel, P. (2008). *Ethics in psychology and the mental health professions: Standards and cases* (3rd ed.). New York: Oxford University Press.

Koopmans, M. (2009). Epilogue: Psychology at the edge of chaos. In S. J. Guastello, M. Koopmans, & D. Pincus (Eds.), *Chaos and complexity in psychology: The theory of nonlinear dynamical systems* (pp. 506–527). New York: Cambridge University Press.

Kosmin, B. A., & Keysar, A. (2009). *American religious identification survey 2008*. Hartford, CN: Institute for the Study of Secularism in Society and Culture.

Kroenke, K., Spitzer, R. L., & Williams, J. B. (2001). The PHQ-9: Validity of a brief depression severity measure. *Journal of General and Internal Medicine*, 16, 603–613.

Krupnick, J. L., Stotsky, S. M., Simmons, S., Moyer, J., Watkins, J., & Elkin, I., et al. (1996). The role of the therapeutic alliance in psychotherapy and pharmacotherapy outcome: Findings in the National Institute of Mental Health Treatment of Depression Collaborative Research Program. *Journal of Consulting and Clinical Psychology*, 64, 532−539.

Kuhn, T. S. (1962). *The structure of scientific revolutions*. Chicago: University of Chicago Press.

Kurzweil, R. (2005). *The singularity is near*. New York: Penguin.

Lahey, B. B. (2009). Public health significance of neuroticism. *American Psychologist*, 64, 241−256.

Lambert, M. J. (1992). Implications of outcome research for psychotherapy integration. In J. C. Norcross, & M. R. Goldstein (Eds.), *Handbook of psychotherapy integration* (pp. 94−129). New York: Basic Books.

Lambert, M. J. (2007). Presidential address: What we have learned from a decade of research aimed at improving psychotherapy outcome in routine care. *Psychotherapy Research*, 17, 1−14.

Lambert, M. J., & Archer, A. (2006). Research findings on the effects of psychotherapy and their implications for practice. In C. D. Goodheart, A. E. Kazdin, & R. J. Sternberg (Eds.), *Evidence-based psychotherapy: Where practice and research meet* (pp. 111−130). Washington, DC: American Psychological Association.

Lambert, M. J., & Barley, D. E. (2002). Research summary on the therapeutic relationship and psychotherapy outcome. In J. C. Norcross (Ed.), *Psychotherapy relationships that work* (pp. 17−32). New York: Oxford University Press.

Lambert, M. J., & Bergin, A. E. (1994). The effectiveness of psychotherapy. In A. E. Bergin, & S. L. Garfield (Eds.), *Handbook of psychotherapy and behavior change* (4th ed., pp. 143−189). New York: Wiley.

Lambert, M. J., Burlingame, G. M., Umphress, V. J., Hansen, N. B., Vermeersch, D., & Clouse, G., et al. (1996). The reliability and validity of the Outcome Questionnaire. *Clinical Psychology and Psychotherapy*, 3, 106−116.

Lambert, M. J., & Ogles, B. M. (2004). The efficacy and effectiveness of psychotherapy. In M. J . Lambert (Ed.), *Garfield and Bergin's handbook of psychotherapy and behavior change: An empirical analysis* (5th ed., pp. 139−193). New York: Wiley.

Lambert, M. J., Ogles, B. M., & Masters, K. S. (1992). Choosing outcome assessment devices: An organizational and conceptual scheme. *Journal of Counseling and Development*, 70, 527−532.

Larson, D. (1980). Therapeutic schools, styles, and schoolism: A national survey. *Journal of Humanistic Psychology*, 20, 3−20.

Last, J. M., & Wallace, R. B. (Eds.), (1992). *Maxcy-Rosenau-Last public health and preventive medicine* (13th ed.). Norwalk, CT: Appleton & Lange.

Lesk, M. (2004). Online data and scientific progress: Content in cyberinfrastructure. Paper presented at Digital Curation Centre, Edinburgh, Scotland, 24 September, 2004. Retrieved March 19, 2007 from http://archiv.twoday.net/stories/337419.

LeVine, E. S. (2007). Experiences from the frontline: Prescribing in New Mexico. *Psychological Services*, 4, 59−71.

Levy, S. (1992). *Artificial life: A report from the frontier where computers meet biology*. New York, NY: Vintage Books.

Lezak, M. D. (1995). *Neuropsychological assessment* (3rd ed.). New York: Oxford University Press.

Lichtenberg, J. W., Partnoy, S. M., Bebeau, M. J., Leigh, I. W., Nelson, P. D., & Rubin, N. J., et al. (2007). Challenges to the assessment of competence and competencies. *Professional Psychology: Research and Practice*, 38, 474−478.

Lilienfeld, S. O. (2007). Psychological treatments that cause harm. *Perspectives on Psychological Science*, 2, 53–70.

Lindsay, R. A. (2005). Slaves, embryos, and nonhuman animals: Moral status and the limitations of common morality theory. *Kennedy Institute of Ethics Journal*, 15, 323–346.

Linehan, M. M., & Dexter-Mazza, E. T. (2008). Dialectical behavior therapy for borderline personality disorder. In D. H. Barlow (Ed.), *Clinical handbook of psychological disorders: A step-by-step treatment manual* (pp. 365–420). New York: Guilford.

Lipsey, M. W., & Wilson, D. B. (1993). The efficacy of psychological, educational, and behavior treatment: Confirmation from meta-analysis. *American Psychologist*, 48, 1181–1209.

Locke, J. (1690). *An essay concerning human understanding.* Retrieved from http://www.gutenberg.org/ebooks/10615

Loeppke, R., Taitel, M., Haufle, V., Parry, T., Kessler, R. C., & Jinnett, K. (2009). Health and productivity as a business strategy: A multiemployer study. *Journal of Occupational and Environmental Medicine*, 51, 411–428.

Loftus, E. F., & Davis, D. (2006). Recovered memories. *Annual Review of Clinical Psychology*, 2, 469–498.

Lorenz, E. N. (1963). Deterministic nonperiodic flow. *Journal of Atmospheric Science*, 357, 130–141.

Loue, S. (2000). *Textbook of research ethics: Theory and practice.* New York: Springer.

Lovelock, J. (1979). *Gaia: A new look at life on earth.* New York: Oxford University Press.

Lumsden, C. J. (2005). The next synthesis: 25 years of genes, mind, and culture. In C. J. Lumsden, & E. O. Wilson (Eds.), *Genes, mind, and culture: the coevolutionary process* (pp. i–xviii). London: World Scientific Publishing.

Lumsden, C. J., & Wilson, E. O. (1981). *Genes, mind, and culture: the coevolutionary process.* London: World Scientific Publishing.

Lunkenheimer, E. S., & Dishion, T. J. (2009). Developmental psychopathology: Maladaptive and adaptive attractors in children's close relationships. In S. J. Guastello, M. Koopmans, & D. Pincus (Eds.), *Chaos and complexity in psychology: The theory of nonlinear dynamical systems* (pp. 282–306). New York: Cambridge University Press.

Maciejewski, M. L. (2006). Generic measures. In R. L. Kane (Ed.), *Understanding health care outcomes research* (2nd ed., pp. 123–164). Sudbury, MA: Jones and Barlett Publishers.

MacIntyre, A. (1982). *After virtue: A study in moral theory.* Notre Dame, IN: University of Notre Dame Press.

Magnavita, J. J. (2005). *Personality-guided relational psychotherapy: A unified approach.* Washington, DC: American Psychological Association.

Magnavita, J. J. (2006). In search of the unifying principles of psychotherapy: Conceptual, empirical, and clinical convergence. *American Psychologist*, 61, 882–892.

Magnavita, J. J. (2008). Toward unification of clinical science: The next wave in the evolution of psychotherapy? *Journal of Psychotherapy Integration*, 18, 264–291.

Main, M., Kaplan, N., & Cassidy, J. (1985). Security in infancy, childhood and adulthood: A move to the level of representation. In I. Bretherton, & E. Waters (Eds.), *Growing points of attachment theory and research.* Chicago: University of Chicago Press.

Malone, M. S. (Nov. 27, 2000). God, Stephen Wolfram, and everything else. *Forbes ASAP.* Retrieved from http://members.forbes.com/asap/2000/1127/162.html

Maloney, M. P., & Ward, M. P. (1976). *Psychological assessment: A conceptual approach.* New York: Oxford University Press.

Malouff, J. M., Thorsteinsson, E. B., & Schutte, N. S. (2005). The relationship between the five-factor model of personality and symptoms of clinical disorders: A meta-analysis. *Journal of Psychopathology and Behavioral Assessment*, 27, 101–114.

Manning, M., Homel, R., & Smith, C. (2010). A meta-analysis of the effects of early developmental prevention programs in at-risk populations on non-health outcomes in adolescence. *Children and Youth Services Review*, 32, 506–519. doi:10.1016/j.childyouth.2009.11.003.

Marks, I. M., & Matthews, A. M. (1978). Brief standard self-rating for phobic patients. *Behavior Research and Therapy*, 17, 263–267.

Martin, P. R., Weinberg, B. A., & Bealer, B. K. (2007). *Healing addiction: An integrated pharmacopsychosocial approach to treatment*. Hoboken, NJ: Wiley.

Maruish, M. E. (2004a). Introduction. In M. E. Maruish (Ed.), *The use of psychological testing for treatment planning and outcomes assessment* (3rd ed., pp. 1–64). Mahwah, NJ: Lawrence Erlbaum.

Maruish, M. E. (2004b). Development and implementation of a behavioral health outcomes program. In M. E. Maruish (Ed.), *The use of psychological testing for treatment planning and outcomes assessment* (3rd ed., pp. 215–272). Mahwah, NJ: Lawrence Erlbaum.

Maslow, A. H. (1943). A theory of human motivation. *Psychological Review*, 50, 370–396.

Mayou, R. A., Ehlers, A., & Hobbs, M. (2000). Psychological debriefing for road traffic accident victims: Three-year follow-up of randomized controlled trial. *British Journal of Psychiatry*, 176, 589–593.

McAdams, D. P., & Pals, J. L. (2006). A new Big Five: Fundamental principles for an integrative science of personality. *American Psychologist*, 61, 204–217.

McClain, T., O'Sullivan, P. S., & Clardy, J. A. (2004). Biopsychosocial formulation: Recognizing educational shortcomings. *Academic Psychiatry*, 28, 88–94.

McGauhey, P., Starfield, B., Alexander, C., & Ensminger, M. E. (1991). Social environmental and vulnerability of low birth weight children: A social-epidemiological perspective. *Pediatrics*, 88, 943–953.

McHugh, R. K., & Barlow, D. H. (2010). The dissemination and implementation of evidence-based psychological treatments: A review of current efforts. *American Psychologist*, 65, 73–84.

McKeown, T. (1979). *The role of medicine: Dream or nemesis?* Oxford, UK: Blackwell.

McLellan, A. T., Kushner, H., Metzger, D., Peters, R., Smith, I., & Grissom, G., et al. (1992). The fifth edition of the Addiction Severity Index. *Journal of Substance Abuse Treatment*, 9, 199–213.

McNeill, W. H. (1979). *Plagues and peoples*. Harmondsworth, UK: Penguin.

Meier, S. T., & Davis, S. R. (1990). Trends in reporting psychometric properties of scales used in counseling psychology research. *Journal of Counseling Psychology*, 37, 113–115.

Melchert, T. P. (2007). Strengthening the scientific foundations of professional psychology: Time for the next steps. *Professional Psychology: Research and Practice*, 38, 34–43.

Merikangas, K. R., Ames, M., Cui, L., Stang, P. E., Ustun, T. B., & Von Korff, M., et al. (2007). The impact of comorbidity of mental and physical conditions on role disability in the US adult household population. *Archives of General Psychiatry*, 64, 1180–1188.

Messer, S. B. (2004). Evidence-based practice: Beyond empirically supported treatments. *Professional Psychology*, 35, 580–588.

Meyer, G. J., Finn, S. E., Eyde, L. D., Kay, G. G., Moreland, K. L., & Dies, R. R., et al. (2001). Psychological testing and psychological assessment: A review of evidence and issues. *American Psychologist*, 56, 128–165.

Meyer, L. (2008). The use of a comprehensive biopsychosocial framework for intake assessment in mental health practice. Doctoral dissertation, Marquette University, Milwaukee, WI. Retrieved from http://0-proquest.umi.com.libus.csd.mu.edu/pqdweb?did=1592080121&Fmt=7&clientId=1953&RQT=309&VName = PQD

Meyer, L., & Melchert, T. P. (2011). Examining the content of mental health intake assessments from a biopsychosocial perspective. *Journal of Psychotherapy Integration*, 21, 70–89.

Mikulincer, M., & Shaver, P. R. (2007). *Attachment in adulthood: Structure, dynamics, and change*. New York: Guilford.

Miller, R. C., & Berman, J. S. (1983). The efficacy of cognitive behavior therapies: A quantitative review of the research evidence. *Psychological Bulletin*, 94, 39–53.

Minelli, T. A. (2009). Neurodynamics and electrocortical activity. In S. J. Guastello, M. Koopmans, & D. Pincus (Eds.), *Chaos and complexity in psychology: The theory of nonlinear dynamical systems* (pp. 73–107). New York: Cambridge University Press.

Mitchell, M. (2009). *Complexity: A guided tour*. New York: Oxford University Press.

Mokdad, A. H., Marks, J. S., Stroup, D. F., & Gerberding, J. L. (2004). Actual causes of death in the United States, 2000. *Journal of the American Medical Association*, 291, 1238–1245.

Moore, G. E. (1903). *Principia Ethica*. Cambridge, UK: Cambridge University Press.

Mrazek, P. J., & Hall, M. (1997). A policy perspective on prevention. *American Journal of Community Psychology*, 25, 221–226.

Munoz, R. F., Hollon, S. D., McGrath, E., Rehm, L. P., & VandenBos, G. R. (1994). On the AHCPR depression in primary care guidelines: Further considerations for practitioners. *American Psychologist*, 49, 42–61.

Murray, C. J. L., & Lopez, A. D. (Eds.), (1996). *The global burden of disease: A comprehensive assessment of mortality and disability from diseases, injuries, and risk factors in 1990 and projected to 2020* Cambridge, MA: Harvard School of Public Health.

Myers, J. E. B., Berliner, L., Briere, J., Hendrix, C. T., Jenny, C., & Reid, T. A. (2002). *The APSAC handbook on child maltreatment* (2nd ed.). Thousand Oaks, CA: Sage.

National Association of Social Workers. (2008). *Code of ethics*. Retrieved from http://www.socialworkers.org/pubs/code/code.asp

National Center for Educational Statistics. (2008). *National assessment of adult literacy*. Retrieved from nces.ed/gov/naal/estimates/overview.aspx

National Council of Schools of Professional Psychology. (2007). *Competency Developmental Achievement Levels (DALs) of the National Council of Schools and Programs in Professional Psychology (NCSPP)*. Retrieved from http://www.ncspp.info/DALof%20NCSPP%209-21-07.pdf

National Institute of Mental Health. (1993). *The prevention of mental disorders: A national research agenda*. Rockville, MD: Author.

National Institute of Mental Health. (1995). *A plan for prevention research for the National Institute of Mental Health: A report to the National Advisory Mental Health Council*. Rockville, MD: Author.

National Research Council and Institute of Medicine. (2009). *Preventing mental, emotional and behavioral disorders among young people: Progress and possibilities*. Washington, DC: National Academies Press.

Nelson, P. D. (2007). Striving for competence in the assessment of competence: Psychology's professional education and credentialing journey of public accountability. *Training and Education in Professional Psychology*, 1, 3–12.

New Freedom Commission on Mental Health. (2003). *Achieving the promise: Transforming mental health care in America. Final Report* [DHHS Pub. No. SMA-03-3832]. Rockville, MD: U.S. Department of Health and Human Services.

Newsweek. (2005). *Newsweek/Beliefnet survey*. Retrieved from www.beliefnet.com/story/173/story_17353.html

Nichols, J. O., & Nichols, K. W. (2001). *General education assessment for improvement of student academic achievement: Guidance for academic departments and committees.* New York: Agathon.

Nichols, J. P. (1998). *Family therapy: concepts and methods.* Boston: Allyn & Bacon.

Nietzel, M. T., Russell, R. L., Hemmings, K. A., & Gretter, M. L. (1987). Clinical significance of psychotherapy for unipolar depression: A meta-analytic approach to social comparison. *Journal of Consulting and Clinical Psychology, 55,* 156–161.

Norcross, J. C. (Ed.), (2002). *Psychotherapy relationships that work: Therapist contributions and responsiveness to patients* New York: Oxford University Press.

Norcross, J. C. (2005). A primer on psychotherapy integration. In J. C. Norcross, & M. R. Goldfried (Eds.), *Handbook of psychotherapy integration* (2nd ed., pp. 3–23). New York: Oxford University Press.

Norcross, J. C., Beutler, L. E., & Levant, R. F. (Eds.), (2006). *Evidence-based practices in mental health: Debate and dialogue on the fundamental questions* Washington, DC: American Psychological Association.

Norcross, J. C., & Goldfried, M. R. (2005). *Handbook of psychotherapy integration* (2nd ed.). New York: Oxford University Press.

Norcross, J. C., & Lambert, M. J. (2006). The therapy relationship. In J. C. Norcross, L. E. Beutler, & R. F. Levant (Eds.), *Evidence-based practices in mental health: Debate and dialogue on the fundamental questions* (pp. 208–217). Washington, DC: American Psychological Association.

Nyman, J. A. (2006). Cost-effectiveness analysis. In R. L. Kane (Ed.), *Understanding health care outcomes research* (2nd ed., pp. 335–349). Sudbury, MA: Jones and Barlett Publishers.

Ogles, B. M., Lambert, M. J., & Fields, S. A. (2002). *Essentials of outcome assessment.* New York: John Wiley & Sons.

Ogles, B. M., Lambert, M. J., & Masters, K. S. (1996). *Assessing outcome in clinical practice.* Boston: Allyn & Bacon.

Okiishi, J., Lambert, M. J., Nielsen, S. L., & Ogles, B. M. (2003). In search of supershrink: Using patient outcome to identify effective and ineffective therapists. *Clinical Psychology and Psychotherapy, 10,* 361–373.

Okiishi, J. C., Lambert, M. J., Eggett, D., Nielsen, S. L., Dayton, D. D., & Vermeersch, D. A. (2006). An analysis of therapist treatment effects: Toward providing feedback to individual therapists on their patients' psychotherapy outcome. *Journal of Clinical Psychology, 62,* 1157–1172.

Ollendick, T. (2008). Foreward. In R. T. Brown, D. O. Antonuccio, G. J. DuPaul, M. A. Fristad, C. A. King, & L. K. Leslie, et al. *Childhood mental health disorders: Evidence base and contextual factors for psychosocial, psychopharmacological, and combined interventions* (pp. vii–ix). Washington, DC: American Psychological Association.

Ollendick, T. H., & King, N. J. (2006). Empirically supported treatments typically produce outcomes superior to non-empirically supported treatment therapies. In J. C. Norcross, L. E. Beutler, & R. F. Levant (Eds.), *Evidence-based practices in mental health: Debate and dialogue on the fundamental questions* (pp. 308–317). Washington, DC: American Psychological Association.

O'Reilly, R. C. (2006). Biologically based computational models of high-level cognition. *Science, 314,* 91–94.

Otto, M. W., Smits, J. A. J., & Reese, H. E. (2005). Combined psychotherapy and pharmacotherapy for mood and anxiety disorders in adults: Review and analysis. *Clinical Psychology: Science and Practice, 12,* 72–86.

Partnership for Solutions. (2004). *Chronic conditions: Making the case for ongoing care.* Baltimore, MD: Johns Hopkins University Press.

Patrick, D. L., & Deyo, R. A. (1989). Generic and disease-specific measures in assessing health status and quality of life. *Medical Care*, 27, 5217–5232.

Paykel, E. S., Scott, J., Teasdale, J. D., Johnson, A. L., Garland, A., & Moore, R., et al. (1999). Prevention of relapse in residual depression by cognitive therapy: A controlled trial. *Archives of General Psychiatry*, 56, 829–835.

Perkins, B. R., & Rouanzoin, C. C. (2002). A critical evaluation of current views regarding eye movement desensitization and reprocessing (EMDR): Clarifying points of confusion. *Journal of Clinical Psychology*, 58, 77–97.

Peterson, R. L., McHolland, J. D., Bent, R. J., Davis-Russell, E., Edwall, G. E., & Polite, K., et al. *The core curriculum in professional psychology.* Washington, DC: American Psychological Association.

Pincus, D. (2009). Coherence, complexity and information flow: Self-organizing processes in psychotherapy. In S. J. Guastello, M. Koopmans, & D. Pincus (Eds.), *Chaos and complexity in psychology: The theory of nonlinear dynamical systems* (pp. 335–369). New York: Cambridge University Press.

Pinker, S. (2002). *The blank slate: The modern denial of human nature.* New York: Viking.

Poincare, H. (1914). Science and method *(Francis Maitland, Trans.).* London: Nelson & Sons.

Pollan, M. (2006). *The omnivore's dilemma: The natural history of four meals.* London: Penguin.

Ponterotto, J. G., Casas, M., Suzuki, L. A., & Alexander, C. M. (2001). *Handbook of multicultural counseling* (2nd ed.). Thousand Oaks, CA: Sage.

Popper, K. (1963). *Conjectures and refutations: The growth of scientific knowledge.* London: Routledge.

Prigogine, I., & Stengers, I. (1984). *Order out of chaos: Man's new dialog with nature.* New York: Bantam.

Prilleltensky, I. (2008). The role of power in wellness, oppression, and liberation: The promise of psychopolitical validity. *Journal of Community Psychology*, 36, 116–136.

Prior, V., & Glaser, D. (2006). *Understanding attachment and attachment disorders: Theory, evidence, and practice.* London: Jessica Kingsley Publishers.

Prochaska, J. O., & Norcross, J. C. (2010). *Systems of psychotherapy: A Transtheoretical analysis* (7th ed.). Belmont, CA: Thompson Brooks/Cole.

Project Release v. Prevost, 722 F.2d 960 (2nd Cir., 1983).

Raimy, V. C. (Ed.), (1950). *Training in clinical psychology.* Englewood Cliffs, NJ: Prentice-Hall.

Rawls, J. (1999). *A theory of justice* (2nd ed.). Cambridge, MA: Harvard University Press.

Reed, G. M., & Eisman, E. J. (2006). Uses and misuses of evidence: Managed care, treatment guidelines, and outcomes measure in professional practice. In C. D. Goodheart, A. E. Kazdin, & R. J. Sternberg (Eds.), *Evidence-based psychotherapy: Where practice and research meet* (pp. 13–36). Washington, DC: American Psychological Association.

Renaud, J., Brent, D. A., Baugher, M., Birmaher, B., Kolko, D. J., & Bridge, J. (1998). Rapid response to psychosocial treatment for adolescent depression: A two-year follow-up. *Journal of the American Academy of Child and Adolescent Psychiatry*, 37, 1184–1191.

Rescher, N. (1966). *Distributive justice.* Indianapolis, IN: Bobbs-Merrill.

Robins, L. N. (1970). Follow-up studies investigating childhood disorders. In E. H. Hare, & J. K. Wayne (Eds.), *Psychiatric epidemiology* (pp. 29–68). London: Oxford University Press.

Robinson, L. A., Berman, J. S., & Neimeyer, R. A. (1990). Psychotherapy for the treatment of depression: A comprehensive review of controlled outcome research. *Psychological Bulletin*, 108, 30–49.

Rogers, C. R. (1961). *On becoming a person*. Boston: Houghton Mifflin.

Rogers, J. L. (2010). The epistemology of mathematical and statistical modeling: A quiet methodological revolution. *American Psychologist*, 65, 1–12.

Rosenzweig, S. (1936). Some implicit common factors in diverse methods of psychotherapy: "At last the Dodo said, 'Everybody has won and all must have prizes.'". *American Journal of Orthopsychiatry*, 6, 412–415.

Rothstein, H. R., Sutton, A. J., & Borenstein, M. (Eds.), (2005). *Publication bias in meta-analysis: Prevention, assessment and adjustments* Chichester, UK: John Wiley & Sons.

Rubin, N. J., Bebeau, M., Leigh, I. W., Lichtenberg, J. W., Nelson, P. D., & Portnoy, S., et al. (2007). The competency movement within psychology: An historical perspective. *Professional Psychology: Research and Practice*, 38, 452–462.

Rudd, M. D. (2006). *The assessment and management of suicidality*. Sarasota, FL: Professional Resource Press.

Rutter, M. (1979). Protective factors in children's responses to stress and disadvantage. *Annals of the Academy of Medicine, Singapore*, 8, 324–338.

Rychlak, J. R. (2005). Unification in theory and method: Possibilities and impossibilities. In R. J. Sternberg (Ed.), *Unity in psychology: Possibility or pipedream?* (pp. 145–158). Washington, DC: American Psychological Association.

Sandel, M. (2005). *Democracy's discontent: America in search of a public philosophy*. Cambridge, MA: Harvard University Press.

Saulsman, L. M., & Page, A. C. (2004). The five-factor model and personality disorder empirical literature: A meta-analytic review. *Clinical Psychology Review*, 23, 1055–1085.

Schiff, G. D., Hasan, O., Kim, S., Abrams, R., Cosby, K, & Lambert, B. L., et al. (2009). Diagnostic error in medicine. *Archives of Internal Medicine*, 169, 1881–1887.

Schone-Seifert, B. (2006). Danger and merits of principlism: Meta-theoretical reflections on the Beauchamp/Childress approach to biomedical ethics. In C. Rehmann-Sutter, M. Duwell, & D. Mieth (Eds.), *Bioethics on cultural contexts: Reflections on methods and finitude* (pp. 109–119). Dordrecht, Netherlands: Springer.

Schwartz, J. M., & Begley, S. (2002). *The mind and the brain: Neuroplasticity and the power of mental force*. New York: HarperCollins.

Seagull, E. A. (2000). Beyond mothers and children: Finding the family in pediatric psychology. *Journal of Pediatric Psychology*, 25, 161–169.

Seligman, L., & Reichenberg, L. W. (2007). *Selecting effective treatments: A comprehensive, systematic guide to treating mental disorders* (3rd ed.). San Francisco: Jossey-Bass.

Seligman, M. E. P., Rashid, T., & Parks, A. C. (2006). Positive psychotherapy. *American Psychologist*, 61, 774–788.

Shah, S., & Reichman, W. E. (2006). Psychiatric intervention in long-term care. In L. Hyer, & R. C. Intrieri (Eds.), *Geropsychological interventions in long-term care* (pp. 85–107). New York: Springer.

Sharpe, V. A., & Faden, A. I. (1998). *Medical harm: Historical, conceptual, and ethical dimensions of iatrogenic illness*. Cambridge, UK: Cambridge University Press.

Shedler, J. (2010). The efficacy of psychodynamic psychotherapy. *American Psychologist*, 65, 98–109.

Simon, H. (1962). The architecture of complexity. *Proceedings of the American Philosophical Society*, 106, 467–482.

Smith, A. (1776). *The wealth of nations*. Retrieved from http://www.gutenberg.org/etext/ 3300

Smith, M. A., Schussler-Fiorenza, C., & Rockwood, T. (2006). Satisfaction with care. In R. L. Kane (Ed.), *Understanding health care outcomes research* (2nd ed., pp. 185–218). Sudbury, MA: Jones and Barlett.

Smith, M. B. (2000). Moral foundations in research with human participants. In B. D. Sales, & S. Folkman (Eds.), *Ethics in research with human participants* (pp. 3–10). Washington, DC: American Psychological Association.

Smith, M. L., & Glass, G. V. (1977). Meta-analysis of psychotherapy outcome studies. *American Psychologist*, 32, 752–760.

Smith, M. L., Glass, G. V., & Miller, T. I. (1980). *The benefits of psychotherapy*. Baltimore, MD: Johns Hopkins University Press.

Sobell, M. B., & Sobell, L. C. (2000). Stepped care as a heuristic approach to the treatment of alcohol problems. *Journal of Consulting and Clinical Psychology*, 68, 573–579.

Sowers, W., George, C., & Thompson, K. (1999). Level of Care Utilization System for Psychiatric and Addiction Services (LOCUS): A preliminary assessment of reliability and validity. *Community Mental Health Journal*, 35, 545–563.

Sowers, W., Pumariega, A., Huffine, C., & Fallon, T. (2003). Best practices: Level-of-care decision making in behavioral health services: The LOCUS and the CALOCUS. *Psychiatric Services*, 54, 1461–1463.

Spanier, G. B. (1976). Measuring dyadic adjustment: New scales for assessing the quality of marriage and similar dyads. *Journal of Marriage and the Family*, 38, 15–28.

Speer, D. C. (1998). *Mental health outcome evaluation*. San Diego, CA: Academic Press.

Sperry, L. (1988). Biopsychosocial therapy: An integrative approach for tailoring treatment. *Individual Psychology*, 44, 225–235.

Sperry, L. (1999). Biopsychosocial therapy. *The Journal of Individual Psychology*, 55, 233–247.

Sperry, L. (2001). The biological dimension on biopsychosocial therapy: Theory and clinical applications with couples. *The Journal of Individual Psychology*, 57, 310–317.

Sperry, L. (2006). *Psychological treatment of chronic illness: The biopsychosocial therapy approach*. Washington, DC: American Psychological Association.

Sperry, L., Brill, P. L., Howard, K. I., & Grissom, G. R. (1996). *Treatment outcomes in psychotherapy and psychiatric interventions*. New York: Brunner/Mazel.

Spielberger, C. D., Gorsuch, R. L., & Lushene, R. E. (1970). *Manual for the State-Trait Anxiety Inventory*. Palo Alto, CA: Consulting Psychologists Press.

Spitzer, R. L., Kroenke, R., Williams, J. B., & Lowe, B. (2006). A brief measure for assessing generalized anxiety disorder: The GAD-7. *Archives of Internal Medicine*, 166, 1092–1097.

Srebnik, D. S., Uehara, E., Smukler, M., Russo, J. E., Comtois, K. A., & Snowden, M. (2002). Psychometric properties and utility of the Problem Severity Summary for adults with serious mental illness. *Psychiatric Services*, 53, 1010–1017.

Staats, A. W. (1963). *Complex human behavior*. New York: Holt, Rinehart, & Winston.

Staats, A. W. (1983). *Psychology's crisis of disunity: Philosophy and method for a unified science*. New York: Praeger.

Staats, A. W. (1999). Unifying psychology requires new infrastructure, theory, methods, and research agenda. *Review of General Psychology*, 3, 3–13.

Staats, A. W. (2005). A road to, and philosophy of, unification. In R. J. Sternberg (Ed.), *Unity in psychology: Possibility or pipedream?* (pp. 159–178). Washington, DC: American Psychological Association.

Steele, H., Steele, M., & Fonagy, P. (1996). Associations among attachment classifications of mothers, fathers, and their infants. *Child Development*, 67, 541−555. doi:10.2307/1131831.

Sternberg, R. J. (2005). Unifying the field of psychology. In R. J. Sternberg (Ed.), *Unity in psychology: Possibility or pipedream?* (pp. 3−14). Washington, DC: American Psychological Association.

Sternberg, R. J., & Grigorenko, E. L. (2001). Unified psychology. *American Psychologist*, 56, 1069−1079.

Stewart, I. (1989). *Does God play dice?* Cambridge, MA: Blackwell.

Stewart, I. (1995). *Concepts of modern mathematics*. Mineola, NY: Dover Publications.

Stromberg, C. D., Haggarty, D. J., Mishkin, B., Leibenluft, R. F., Rubin, B. L., & McMillian, M. H., et al. *The psychologist's legal handbook*. Washington, DC: Council for the National Register of Health Service Providers in Psychology.

Strum, P. (1999). *When the Nazi's came to town: Freedom for speech we hate*. Lawrence, KS: University of Kansas Press.

Strupp, H. H., & Hadley, S. M. (1977). A tripartite model of mental health and therapeutic outcomes. *American Psychologist*, 32, 187−196.

Sue, D. W., & Sue, D. (1990). *Counseling the culturally diverse: Theory and practice*. New York: Wiley.

Sue, D. W., & Sue, D. (2008). *Counseling the culturally diverse: Theory and practice* (5th ed.). New York: Wiley.

Sulis, W. (2009). Collective intelligence: Observations and models. In S. J. Guastello, M. Koopmans, & D. Pincus (Eds.), *Chaos and complexity in psychology: The theory of nonlinear dynamical systems* (pp. 41−72). New York: Cambridge University Press.

Suls, J., & Rothman, A. (2004). Evolution of the biopsychosocial model: Prospects and challenges for health psychology. *Health Psychology*, 23, 119−125.

Suzuki, L. A., Meller, P. J., & Ponterotto, J. G. (2008). *Handbook of multicultural assessment: Clinical, psychological and educational implications* (3rd ed.). San Francisco: Jossey-Bass.

Szasz, T. S. (1971). The sane slave: An historical note on the use of medical diagnosis as justificatory rhetoric. *American Journal of Psychotherapy*, 25, 228−239.

Tallent, N. (1992). *The practice of psychological assessment*. Englewood Cliffs, NJ: Prentice-Hall.

Tarasoff v. Board or Regents of the University of California (1976). 17 Cal. 3d 425, 551 P.2d 334, 131 Cal Reptr. 14.

Taylor, S., & McLean, P. (1993). Outcome profiles in the treatment of unipolar depression. *Behavioral Research and Therapy*, 31, 325−330.

Teasdale, J. D., Segal, Z. V., Williams, J. M. G., Ridgeway, V. A., Soulsby, J. M., & Lau, M. A. (2000). Prevention of relapse/recurrence in major depression by mindfulness-based cognitive therapy. *Journal of Consulting and Clinical Psychology*, 68, 615−623.

Thase, M. E., & Jindal, R. D. (2004). Combining psychotherapy and psychopharmacology for treatment of mental disorders. In M. J . Lambert (Ed.), *Garfield and Bergin's handbook of psychotherapy and behavior change: An empirical analysis* (5th ed., pp. 743−766). New York: Wiley.

Tobler, N. S., & Stratton, H. H. (1997). Effectiveness of school-based drug prevention programs: A meta-analysis of the research. *Journal of Primary Prevention*, Vol 18, 71−128.

Toro, P. A., Tompsett, C. J., Lombardo, S., Philippot, P., Nachtergael, H., & Galand, B., et al. (2007). Homelessness in Europe and the United States: A comparison of prevalence and public opinion. *Journal of Social Issues*, 63, 505−524.

Trull, T. J., Nietzel, M. T., & Main, A. (1988). The use of meta-analysis to assess the clinical significance of behavior therapy for agoraphobia. *Behavior Therapy*, 19, 527–538.

Turner, E. H., Matthews, A. M., Linardatos, E., Tell, R. A., & Rosenthal, R. (2008). Selective publication of antidepressant trials and its influence on apparent efficacy. *New England Journal of Medicine*, 358, 252–260.

Turner, S. M., DeMers, S. T., Fox, H. R., & Reed, G. M. (2001). Guidelines for test user qualifications: An executive summary. *American Psychologist*, 56, 1099–1113.

United Nations. (1948). *Universal declaration of human rights of the United Nations.* New York: United Nations.

U.S. Bureau of the Census. (2000). *Census 2000.* Retrieved from censtats.census.gov/data/US/01000.pdf

U.S. Bureau of the Census. (2008). *An older and more diverse nation by midcentury.* Retrieved from http://www.census.gov/PressRelease/www/releases/archives/population/012496.html

U.S. Bureau of Justice Statistics. (2007). *Key crime and justice facts at a glance.* Retrieved from http://www.ojp.usdoj.gov/bjs/glance.htm#Crime

U.S. Department of Health and Human Services. (1999). *Mental health: A report of the Surgeon General.* Rockville, MD: Department of Health and Human Services, Substance Abuse and Mental Health Services Administration, Center for Mental Health Services, National Institute of Mental Health.

U.S. Department of Health and Human Services, Administration on Children, Youth and Families. (2008). *Child maltreatment 2006.* Washington, DC: U.S. Government Printing Office.

U.S. Substance Abuse and Mental Health Services Administration. (2008). *Screening brief intervention, and referral to treatment.* Retrieved from http://sbirt.samhsa.gov/about.htm

Vallacher, R. R., & Nowak, A. (1994). The chaos in social psychology. In R. R. Vallacher, & A. Nowak (Eds.), *Dynamical systems in social psychology* (pp. 1–16). San Diego, CA: Academic Press.

Vallacher, R. R., & Nowak, A. (2007). Dynamical social psychology: Finding order in the flow of human experience. In A. W. Kruglanski, & E. T. Higgins (Eds.), *Social psychology: Handbook of basic principles* (2nd ed., pp. 734–758). New York: Guilford.

van Geert, P. (2009). Nonlinear complex dydnamical systems in developmental psychology. In S. J. Guastello, M. Koopmans, & D. Pincus (Eds.), *Chaos and complexity in psychology: The theory of nonlinear dynamical systems* (pp. 242–281). New York: Cambridge University Press.

Von Bertalanffy, L. (1950). An outline of general system theory. *The British Journal for the Philosophy of Science*, 1, 134–165.

Von Bertalanffy, L. (1968). *General system theory: Foundations, development, applications.* New York: George Braziller.

Von Foerster, H. (1972). Responsibilities of competence. *Journal of Cybernetics*, 2, 1–6.

Vyse, S. (2008). *Going broke: Why Americans can't hold on to their money.* New York: Oxford University Press.

Wachter, R. M. (2009). Entering the second decade of the patient safety movement: The field matures. *Archives of Internal Medicine*, 169, 1894–1896.

Wahba, A., & Bridgewell, L. (1976). Maslow reconsidered: A review of research on the need hierarchy theory. *Organizational Behavior and Human Performance*, 15, 212–240.

Wallin, D. J. (2007). *Attachment in psychotherapy.* New York: Guilford.

Wampold, B. E. (2001). *The great psychotherapy debate: Models, methods, and findings.* Mahwah, NJ: Lawrence Erlbaum Associates.

Wampold, B. E. (2006). Not a scintilla of evidence to support empirically supported treatemsnt as more effective than other treatments. In J. C. Norcross, L. E. Beutler, & R. F. Levant (Eds.), *Evidence-based practices in mental health: Debate and dialogue on the fundamental questions* (pp. 299−308). Washington, DC: American Psychological Association.

Wampold, B. E. (2007). Psychotherapy: The humanistic (and effective) treatment. *American Psychologist, 62,* 857−873.

Wampold, B. E., & Brown, G. S. (2005). Estimating therapist variability: A naturalistic study of outcomes in managed care. *Journal of Consulting and Clinical Psychology, 73,* 914−923.

Wampold, B. E., Lichtenberg, J. W., & Waehler, C. A. (2002). Principles of empirically supported interventions in counseling psychology. *The Counseling Psychologist, 30,* 197−217.

Wampold, B. E., Mondin, G. W., Moody, M., Stich, F., Benson, K., & Ahn, H. (1997). A meta-analysis of outcome studies comparing bona fide psychotherapies: Empirically, "all must have prizes". *Psychological Bulletin, 122,* 203−215.

Wang, P. S., Demler, O., Olfson, M., Wells, K. B., & Kessler, R. C. (2006). Changing profiles of service sectors used for mental health care in the United States. *American Journal of Psychiatry, 163,* 1187−1198.

Wang, P. S., Lane, M., Olfson, M., Pincus, H. A., Wells, K. B., & Kessler, R. C. (2005). Twelve-month use of mental health services in the United States: Results from the National Comorbidity Survey Replication. *Archives of General Psychiatry, 62,* 629−640.

Ware, J. E., & Sherbourne, C. D. (1992). The MOS 36-item Short-Form Health Survey (SF-36). I. Conceptual framework and item selection. *Medical Care, 30,* 473−483.

Washington, H. A. (2007). *Medical apartheid: A dark history of medical experimentation on black Americans from Colonial Times to the present.* New York: Doubleday.

Waskow, I. E., & Parloff, M. B. (1975). *Psychotherapy change measures.* Rockville, MD: National Institute of Mental Health.

Watson, J. B. (1925). *Behaviorism.* Chicago: University of Chicago Press.

Watts, L. (2003). Visualizing complexity in the brain. In D. Fogel, & C. Robins (Eds.), *Computational intelligence: The experts speak.* Piscataway, NJ: IEEE Press/Wiley.

Weiss, B., Caron, A., Ball, S., Tapp, J., Johnson, M., & Weisz, J. (2005). Iatrogenic effects of group treatment for antisocial youth. *Journal of Consulting and Clinical Psychology, 73,* 1036−1044.

Welfel, E. R. (2010). *Ethics in counseling and psychotherapy: Standards, research, and emerging issues* (4th ed.). Belmont, CA: Brooks/Cole.

Wellner, A. M. (1978). *Proposal for a national commission on education and credentialing in psychology.* Washington, DC: American Psychological Association.

West, G. B., Brown, J. H., & Enquist, B. J. (1999). The fourth dimension of life: Fractal geometry and allometric scaling of organisms. *Science, 284,* 1677−1679.

White, P. (2005). *Biopsychosocial medicine: An integrated approach to understanding illness.* New York: Oxford University Press.

Wiggins, J. S. (2003). *Paradigms of personality assessment.* New York: Guilford.

Wilkinson, L. and the Task Force on Statistical Inference. (1999). Statistical methods in psychology journals: Guidelines and explanations. *American Psychologist, 54,* 594−604. doi:10.1037/0003-066X.54.8.594.

Williams, B. (1994). Patient satisfaction: A valid concept? *Social Sciences and Medicine*, 38, 509–516.

Williams, W. H., & Evans, J. J. (2003). *Biopsychosocial approaches in neurorehabilitation: Assessment and management of neuropsychiatric, mood and behavioural disorders.* Hove, UK: Psychology Press.

Wilson, E. O. (1998). *Consilience: The unity of knowledge.* New York: Knopf.

Wilson, G. T., Vitousek, K. M., & Loeb, K. L. (2000). Stepped care treatment for eating disorders. *Journal of Consulting and Clinical Psychology*, 68, 564–572.

Wolfe, D. (1946). The reorganized American Psychological Association. *American Psychologist*, 1, 3–6.

Wolfram, S. (2002). *A new kind of science.* Champaign, IL: Wolfram Media.

Wolpe, J. (1958). *Psychotherapy by reciprocal inhibition.* Stanford, CA: Stanford University Press.

Wood, C. C., Berger, T. W., Bialek, W., Boahen, K., Brown, E. N., & Holmes, T. C., et al. *Steering Group Report: Brain science as a mutual opportunity for the physical and mathematical sciences, computer science, and engineering.* Arlington, VA: National Science Foundation.

World Health Organization. (1948). *Constitution of the World Health Organization.* Geneva, Switzerland: World Health Organization.

World Health Organization. (2000). *The World Health Report 2000.* Geneva, Switzerland: World Health Organization.

World Health Organization. (2002). *The World Health Report 2002—Reducing risks, promoting health life.* Geneva, Switzerland: World Health Organization.

World Health Organization. (2009). *World Health Statistics 2009.* Geneva, Switzerland: World Health Organization.

Zimmerman, M., Lish, J. D., Lush, D. T., Farber, N. J., Plescia, G., & Kuzma, M. A. (1995). Suicidal ideation among urban medical outpatients. *Journal of General Internal Medicine*, 10, 573–576.

Zoellner, L. A., Feeny, N. C., Cochran, B., & Pruitt, L. (2003). Treatment choice for PTSD. *Behaviour Research and Therapy*, 41, 879–886.

CPSIA information can be obtained at www.ICGtesting.com
Printed in the USA
BVOW09*1620240814

363650BV00002B/12/P